50 IMPORTANT THINGS YOU CAN DO TO IMPROVE EDUCATION

Compiled by Susan Kranberg
for the Book Industry Study Group, Inc.
and the National Association of
Partners in Education, Inc.

Printed in the United States of America

Library of Congress Cataloging-in-Publication Data
Kranberg, Susan, 1950–
    50 important things you can do to improve education / compiled
by Susan Kranberg for Book Industry Study Group and National Association
of Partners in Education
       p.   cm.   Includes bibliographical references and index.
       ISBN 0-940016-41-9
     1. School improvement programs—United States—Directories.   2. Industry and education—United States—Directories.   3. Community and school—United States—Directories.   I. Book Industry Study Group.   II. National Association of Partners in Education.   III. Title.   IV. Title: Fifty important things you can do to improve education.
    LB2822.82.K73   1993   370.19'316'0973—dc20   93-6602   CIP

CONTENTS

Foreword   5

Preface   9

Acknowledgments   13

What Is A Partnership In Education?   17

Introduction   25

SUCCESS STORIES   29

Bantam Doubleday Dell   31

Champion International   34

Houghton Mifflin Company   37

Ingram Distribution Group Inc.   40

Time Warner Inc.   44

EXEMPLARY PROGRAM LISTINGS 47

Adopt-a-School 47

Career Awareness and Leadership
Training 63

Job Readiness 87

Mentoring 99

School Reform and
Curriculum Enhancement 113

Special Awards, Incentives,
and Scholarships 119

Teacher Training 131

Tutoring 153

Resource Directory 171

*National Association of Partners
in Education Contacts by State* 179

Indexes 185

FOREWORD

"Human history becomes more and more a race between education and catastrophe."

H. G. WELLS, *OUTLINE OF HISTORY* (1920)

The health of our society and the strength of our economy depend directly on the success of our system of education, a system that, in turn, cannot succeed without active, innovative support—and sometimes constructive pressure—from all its constituencies. As society's chief consumer of ideas, expertise, and human capital, business is one of the most important of those constituencies. Like any successful relationship, this one involves mutual dependency and requires mutual nurturing.

Let me give a very practical example. Because Time Warner's publishing business depends on the existence of a literate public, we for many years have explored ways to foster literacy—most essentially the ability to read the written word but also an understanding of all the forms of communication that make up our culture, what some people call functional literacy. The most widespread of these efforts is our Time To Read program, described in this volume. We didn't create Time To Read out of some generalized philanthropic impulse or feeling of guilt. We did it, and we continue to develop other similar programs, because they make practical sense for our business.

I believe that every one of the fifty programs detailed in these pages has the same practical foundation. If they didn't, the programs wouldn't have lasted over time or produced concrete results, such as a better educated pool of future employees, or a healthier, less violent community, or a more harmonious business/community relationship.

Perhaps most important for the individual company, well conceived volunteer efforts help produce a more committed work force. In talking with my colleagues who have tutored as part of Time To Read, I have been struck over and over again by how much the experience has meant not just to their students but to *them*—personally, emotionally, and in terms of their deeper appreciation of the impact of ideas, as expressed in both journalistic and creative forms. Their commitment to our company and its mission has broadened and intensified.

But as anyone who has undertaken a serious volunteer effort can testify, along with the enormous personal and practical rewards that come from helping others inevitably comes a sobering, sometimes crushing, awareness of how much more there is to do. Quite frankly, more companies and more individuals need to become involved if our schools and teachers are to have any hope of fulfilling their vital mission.

For that reason, this book performs an invaluable service. By not only suggesting a range of possibilities for business/education partnerships, but also providing the practical information necessary to implement and replicate them, *50 Important Things You Can Do To Improve Education* will,

I hope, stimulate more of our colleagues to join in this desperately important effort. For that, and for the extraordinary programs they themselves have implemented, the members of the Book Industry Study Group deserve all our thanks.

Gerald M. Levin
CHAIRMAN AND CEO, TIME WARNER INC.

PREFACE

The book you are reading is the result of a concern first voiced at a symposium for twenty industry leaders in honor of the fifteenth anniversary of the Book Industry Study Group (BISG) in April 1991.

The study group is an organization that brings together publishers, manufacturers, wholesalers, retailers, librarians, and various trade associations to address issues of common concern and provide research broader in scope than can be undertaken by any single industry association. The group publishes the only annual survey and forecast of book sales (*Book Industry Trends*), the Periodic Consumer Research Study on Book Purchasing, and an array of other publications for the book and serials industry. Two of the most influential of the BISG's working committees are the Book Industry Systems Advisory Committee (BISAC) and the Serials Industry Systems Advisory Committee (SISAC). Both of these play major roles in the streamlining of order and fulfillment systems.

Although the agenda for the symposium did not even

include education, it quickly became clear that the central issue that concerns the book industry as a whole is the critical need for an educated America. Public education is in danger of failing to do the job for a very large contingent of today's youth. Dropout rates are alarming, and the level of knowledge displayed by many graduates of the system is woefully inadequate to enter the work force, let alone to enroll in higher education.

To turn this concern for public education into action, the study group has prepared this book. *50 Important Things You Can Do to Improve Education* is meant to help companies large and small, wholesalers and retailers—in short, anyone in our industry and beyond—to get started with a program of volunteering to improve education in the public schools. Many companies have taken initiatives already, and some of these are described here in some detail. A list of successful programs, sorted by category of activity, follows. Money is by no means the determining factor for involvement; enthusiasm and commitment are.

Our copublisher, the National Association of Partners in Education, Inc. (NAPE), has made its vast data base of partnering programs available to us. NAPE was designated as the National Center for Leadership in Partnerships by the U.S. Department of Education, through a major grant in 1989. This grant—supplemented by private sector funding from NYNEX Corporation and Rockwell International Corporation—supported the creation of NAPE's computerized database of partnerships. Today, AutoZone is principal corporate sponsor of the NAPE National Center.

The NAPE national office and 7,500 grassroots members throughout the country provide leadership in the formation and growth of effective partnerships that ensure success for all students. The organization's principal goals are to increase the number of partnerships nationwide; to promote the need for and benefits of partnerships to policymakers in government, business, and education; and to increase local, state, and national awareness of the importance of partnerships in the success of students.

In the spirit of volunteering, a task force of BISG board members and others worked long and hard to shape an initially general idea into a concrete plan of action. Contributions in kind and in funds from corporations and individuals produced this book; the publicity and distribution are donated as well. Acknowledgements can be found elsewhere; suffice it to say that if the enthusiasm of those who worked to create this book is any measure of the response the book will receive among its readers, we will have reached our goal.

Laura M. H. Conley
BISG CHAIRPERSON

ACKNOWLEDGMENTS

The Book Industry Study Group is very grateful to the following individuals and organizations who have donated time, talent and/or funds to this project. Without their participation, this publication would not exist.

ORGANIZATION CONTRIBUTION

Allen Communications   Publicity, coordination
American Booksellers Association   Publicity
Association of American Publishers   Publicity
AutoZone   NAPE Database Support
Baker & Taylor   Funding
R.R. Donnelley & Sons Company   Printing, binding
Elizabeth Geiser, Consultant   Editor
P.H. Glatfelter Company   Paper
HarperCollins   Funding
Ingram Book Company   Distribution
Macmillan Publishing Company   Design, production

Eileen M. Murphy   Editing
National Association of College Stores   Publicity
New York Public Library   Bibliography
New York State Library   Bibliography
SKP Associates   Indexing
Time Warner   Funding
Waldenbooks   Funding
John Wiley & Sons   Funding

I am grateful that I had the opportunity to participate in this project and discover the important things businesses and individuals are doing in schools across the country. I want to thank everyone who shared information on their education programs for this book and talked to me about their classroom experiences. I drew much encouragement and inspiration from the personal stories that people shared with me and their dedication to children and improving public school education, which I hope is reflected in this book.

I want to thank the Book Industry Study Group for selecting me for this project; Sandy Paul and Bill Raggio from the BISG office for their administrative support; BISG chair, Laura Conley; Paul McLaughlin, George Slowik, Bill Wright and Steve Pekich, who served on the Education Initiative Committee for their feedback at critical times during the development of the book's editorial direction.

I also want to thank the following people for their time and support: at the National Association of Partners in Education, the copublisher of this book, Daniel Merenda, President & CEO; Jane Asche, Director of Development; Linda Tangeman, Director of Field Services, and Susan Otterbourg, President, Delman Educational Communications; Charlotte Frank, Macmillan/McGraw Hill School Division; Jeremiah Kaplan, scholastic; Diana Rigden, Council for Aid to Education; Gail

Niedemhofer and Beth Gabel, U.S. Department of Education, Corporate Liaison Office; Eileen Collins, Time, Inc.; and my friends Alexandra, Kennedy, Bobbi and Marcella, and of course Bob, who supported me when I sat in front of the bland computer screen.

And finally, I want to thank Elizabeth Geiser, the editor of this book. Her excitement about this project was infectious and kept me going when I felt discouraged. And no one could ask for a more sensitive editor. Her care to the details in both content and form made this a far better book. Her generosity of spirit and time served as an example to me that "good" really comes from the doing and giving of service.

Susan Kranberg

WHAT IS A PARTNERSHIP IN EDUCATION?

The National Association of Partners in Education, Inc. (NAPE) defines "partnership in education" as a collaborative effort between a school(s) or school district(s) and one or more community organizations with the purpose of improving the academic and personal growth of America's youth. Businesses, government and community agencies, community clubs and organizations, foundations, colleges and universities, religious organizations, the media, health care agencies, labor organizations, and parent organizations may join in partnership to support school improvement and reform.

A partnership (or collaborative) is formalized with a written agreement or contract which specifies the mutually defined goal(s) and objectives of the partnership, the resources committed, and the activities to be implemented to achieve the stated goal(s).

THE PARTNERSHIPS
IN EDUCATION MOVEMENT

In 1983, the U.S. Department of Education issued *A Nation at Risk—A Report of the National Commission on Excellence in Education*. This highly publicized report has focused America's attention on the critical need for the improvement and reform of our education system. Numerous studies by government agencies, business organizations, and foundation study groups have further defined this need and articulated strategies to meet the need.

Partnerships in education have emerged as one strategy that is making a profound difference in the lives of many American youth. NAPE's recent evaluation of exemplary partnership programs focused on drop-out prevention (Finish for the Future Project) documents their success in improving the academic and employment skills, attitudes, and behaviors of students served by these programs. These outcomes are accomplished through a wide array of programs that include mentoring, tutoring, teacher training, curriculum development, and school restructuring. NAPE estimates that there are now more than 2.6 million volunteers involved in over 200,000 partnerships nationwide.

THE CHANGING PARADIGM
OF PARTNERSHIPS IN EDUCATION

As the body of research on the future of children in America grows, one central theme becomes increasingly clear. If our children are to acquire the citizenship and workforce skills needed to compete in a global economy and live in communities free of poverty, violence, and crime, a comprehensive and integrated system of services (that includes education) must be available to children and their families.

NAPE provides national leadership in bringing growing numbers of citizens, businesses, and other community organizations together to form coalitions that utilize the strategy of partnership to create the community-wide systemic change which is necessary to support effective, long-term school reform efforts. Partnerships have become increasingly sophisticated as partners have come to understand that all segments of the community must work together with the school system to achieve common goals for education improvement and reform.

Many of those involved in education partnerships began with "hands on" or "programmatic" experiences. These initial entry levels of partnership involvement build understanding and trust between educators and their partners from the community at large and pave the way for moving to higher levels of collaboration. Once partners develop a high level of trust in their working relationships, they have laid a sound foundation for taking new steps on an evolving continuum of partnership activities. These new steps often involve the partners in assuming more significant roles in shaping school improvement and reform initiatives. Increasingly, these initiatives are taking the form of collaborative partnerships created for the purpose of providing a comprehensive approach to integrated services for children and families. Such an approach results in community-wide systemic change. Current research provides strong evidence that this change is necessary in order to provide an environment in which all children can learn.

The shift on the part of some partnerships to higher levels of collaboration does not exclude on-going involvement at every level of the partnership continuum. For instance, one-to-one tutoring and mentoring relationships, a "hands on" level of partnership involvement, is an important part of the comprehensive and integrated support system of services which many children presently need to succeed in our schools.

WHAT ARE THE LEVELS OF INVOLVEMENT ALONG THE PARTNERSHIP CONTINUUM?

As a new business partner, your entry level of involvement in a partnership agreement will depend on your organization's past experience with education initiatives and the amount of resources (employee time, money, and materials) you choose to commit. There are four general levels of involvement:

1. **Helping Hands** relationships are those in which an individual company is paired with a school to support, enrich, and improve existing school activities. The company provides the school with goods and services such as tutors, speakers, equipment, and awards and incentives that encourage students to stay in school and strive for significant levels of achievement. Partnerships of this nature are often referred to as adopt-a-school programs.

2. **Programmatic Initiatives** are those in which the individual company works with its school partners to develop programs that target specific curriculum and/or student and teacher needs. Programmatic concerns might include: dropout prevention; attendance; student achievement; technology skills of students and teachers; curriculum development; parent involvement and preparing students for a successful school to work transition.

3. **Policy Changes** are planned and implemented as business leaders, educators and community organizations work together to reduce bureaucratic rules and regulations; advocate proposed policy changes at the legislative level; lobby for

changes in legislation and regulations; and follow through on these efforts to ensure that these changes are supported through adequate financing and programs.

4. **Alliances, Compacts, and Community Coalition Efforts** are joint efforts between several businesses and/or community organizations and one or several schools or school districts. Designed to bring about fundamental educational change, the magnitude of these efforts precludes the support of a single organization. These broad-based collaborative efforts result in action comprehensive enough to produce major changes in the way teaching and learning takes place. Such collaborative efforts produce some of the following outcomes: 1) major changes in curriculum; 2) reorganization of school governance; 3) restructuring of the management and delivery of instruction; 4) improvement in student achievement, behavior and attitudes; 5) community-wide cooperation and coordination in the delivery of comprehensive, integrated services for children and their families and 6) changes in local, state and national policy. This level of partnership activity is by far the most sophisticated and difficult to plan and implement.

THE COLLABORATIVE PROCESS OF PARTNERSHIP DEVELOPMENT

The collaborative process outlined below is designed to provide a "snapshot" of the seven interrelated steps in which your organization should engage with your education partner to ensure an effective partnership in education regardless of your chosen level of entry into the partnership process. Using this generic process to engage in collaborative planning with your chosen education partner will enable you to plan and implement a partnership that can make a significant difference in successful education outcomes for children. Each step is accompanied by a description of the tasks involved in that step.

Getting Started

Map the environment, internal and external to your company, to determine factors that support or hinder the development of a partnership. Create awareness and ownership in the early stages through dialogue among all potential stakeholders in the outcomes of the partnership. Be careful to identify all groups within each partner organization that will see themselves as stakeholders.

Determining Priority Needs and Available Resources

Form a steering committee made up of representatives of the stakeholders. Initiate a process for determining and prioritizing the needs of the school(s) and the business that can be addressed through a partnership without violating the philosophy and values of either organization. Assess the resources currently available among all the partners to meet these needs.

Developing Goals and Objectives

Develop a strategic plan for the partnership including long-range goals and annual objectives that flow from the identified needs and are consistent with the overall philosophy and instructional goals of the school system.

Program Design

Develop an action plan and program materials for achievement of the specific objectives. Define the roles and responsibilities of individuals and partner organizations for implementing the action plan.

Program Management

Develop policies and administrative procedures within the school(s) and the partner organization necessary to ensure that the action plan can be effectively implemented.

Recruiting, Allocating, and Managing Resources

Secure and allocate the necessary material and financial resources. Plan and carry out the recruitment, orientation, training, placement, supervision, and recognition of the volunteers (e.g. the human resources needed to achieve the goals and objectives).

Program Monitoring and Evaluation

Develop and implement a plan to monitor and evaluate both the partnership process and education outcomes for the students to determine if the partnership has been worth the expenditure of time and resources for both the school(s) and the partner organization(s).

For additional information regarding the "how to" of developing, managing, and evaluating partnerships/ collaboratives, contact the National Association of Partners in Education, Inc. (NAPE), 209 Madison Street, Suite 401, Alexandria, Virginia 22314, 703-836-4880.

INTRODUCTION

HOW TO USE THE BOOK

This book has been compiled to stimulate both individual and corporate participation in schools where people live and work. Detailed descriptions of fifty successful programs—spanning a broad range of activities—provide practical answers to such start-up questions as: How do programs originate? Who should manage the program in my company? How much will it cost? Whom can I call for more information? Each listing provides—where available—the name of a contact person who can supply firsthand information.

ORGANIZATION

The main section of the book is organized in nine chapters by major program activity: Adopt-a-School; career awareness and leadership training; job readiness; mentoring; school reform and curriculum enhancement; special awards, incentives and scholarships; teacher training; and tutoring.

Although programs often embrace multiple activities, each is listed under the chapter heading that reflects its major emphasis. The programs are listed alphabetically by program name within each chapter.

Three separate indexes provide access to program details by activity, by company, and by program name. Five anecdotal success stories, based on personal interviews, have been provided to expand upon the factual program description and to highlight the experience of individuals and the corporations they work for in contributing time, energy, and money to improve the lot of public schools and their students.

The appendix consists of two directories that provide sources of additional information. One lists national education abd volunteer organizations and government agencies that offer help to those who want to work in the public schools. The other lists the regional offices of the National Association of Partners in Education (NAPE), which has the mission to help facilitate the establishment of business/school partnerships.

Those seeking additional help can turn to their local community and business organizations such as the Chamber of Commerce, Rotary Club, and parents' organizations. NAPE has contributed an additional section outlining the criteria for establishing a good partnership, along with guidelines for evaluating its effectiveness.

METHODOLOGY AND SELECTION CRITERIA

The U.S. Department of Education, Corporate Liaison Office and NAPE provided information that formed the pool of candidates from which program selections were made. In choosing the fifty programs for inclusion, every effort was made to provide a good balance—by geographic location, by type of activity, and by funding level. As a result, the book lists fifty programs from seventeen states and the District of Columbia. Some programs are supported by company branch offices in small towns and other by corporate headquarters in large metropolitan areas.

Program information was provided by the companies, school districts, and program organizations; details were finalized through telephone interviews.

To be considered for inclusion in this book, programs had to be in place for two years or more. More than twenty-five percent of them have received national and state recognition as exemplary programs, though recognition was not a necessary criterion. Although twenty percent of the programs listed are sponsored by companies that are in the book industry, an effort has been made to have representation from a variety of industries such as manufacturing, travel, insurance, and banking. Since space limitations allowed for the inclusion of only fifty programs, a number of exemplary programs were necessarily omitted. The BISG Education Advisory Committee approved the final selection.

More than twenty-five different program activities are reported involving different levels of company participation and funding. Activities range from painting a school library, summer internships, and special award programs to teacher development, leadership training, and designing a new curriculum. A complete list of program activities represented in the book can be found in the activity index.

SUCCESS
STORIES

BANTAM DOUBLEDAY DELL

ADOPT-A-SCHOOL PROGRAM

It's one thing for a publisher to donate books but quite another to encourage its employees to get involved in an active, personal way with youngsters to develop a love of reading.

A casual conversation at a college reunion on the state of the public school system in New York prompted a not-so-casual response from Dell's executive editor, Emily Reichert. She became fired with the idea that it was time for book publishers to get out of their ivory towers and do something positive in the schools.

She brought her concerns to Isabel Geffner, vice president and associate publisher at Dell, and Craig Virden, vice president of Bantam Doubleday Dell (BDD) Books For Young Readers. Together the three came up with a plan whereby BDD would not only donate books but would also bring their authors and BDD employees into the classroom. They would, in effect, Adopt-A-School.

The proposal was accepted by BDD corporate management and sent to the Mayor's Office of Education Services. "We expected that our plan would be bogged down in red tape," Geffner recalls, "and were shocked when we received a phone call a few days later." A group from BDD met with the mayor's wife, Joyce Dinkins, and her staff to discuss BDD's participation in her citywide Reading Is Recreation read-aloud program founded to encourage first-graders to read. The five schools in the BDD Adopt-A-School program were selected in consultation with Joyce Dinkins and the Mayor's Office for Education Services.

A memo from the BDD chief executive officer, Jack Hoeft, asking for volunteers produced a large response. More than 150 people showed up for the first planning meeting. Sign-up boards gave volunteers the opportunity to choose a school in the borough of their choice, and five school teams were formed.

Before they knew it, this publishing house was definitely out of the ivory tower and knee-deep in children in not one, but five city schools.

"The programs in each of the five schools are very different", Geffner explains. "For example, The Trumpet Book Club, which has adopted P.S. 14 on Staten Island, sponsored a Reading Olympics to promote reading." Medals were awarded to students based on the number of books read, and every student won an award.

"I was delighted to become a team captain at P.S. 48, a school where my mother taught for 25 years," recounts Geffner. "The first thing we did was reopen the school library, which had been closed for ten years. We cleaned, sorted the books, and made new section signs." This year BDD volunteers are working with a fifth grade class at P.S. 48. The class project is a newsletter entitled *Kidz Power*. The first issue included topical articles on Malcolm X, Kwanzaa, and fashion designer Willie

Smith; an article written in Spanish with an English translation; a word-search puzzle; and a stencil to color in.

"The class works on the newsletter three afternoons a week, and we work with them every other week," Geffner reports. "We help them with the editing, layout, and design. When the newsletter is ready, BDD produces it at company headquarters."

With the program in its second year, Geffner admits that maintaining a volunteer pool can be difficult; however, the personal rewards of making a difference in a child's life far outweigh the problems.

"Let's face it," Geffner says, "without these kids experiencing the value of reading, where will we be as publishers in the future?"

Simon & Schuster, HarperCollins, Random House, Scholastic, Macmillan and G.P. Putnam have followed BDD's lead and are now participating in the Reading Is Recreation program.

For complete details on the Adopt-a-School Program, see page 47.

CHAMPION
INTERNATIONAL

MIDDLE SCHOOL PARTNERSHIP PROGRAM

It takes a lot of dedication, patience, and perseverance to get involved with education reform. But the publication in 1989 of *Turning Points: Preparing American Youth for the 21st Century,* the Carnegie Foundation report on adolescent development, brought the issue to the attention of Champion International—and they decided to *do* something about it.

"When deciding to become involved with the education reform issue," recalls Gael Doer, director of corporate contributions and the program's first administrator, "Champion initially wasn't sure where to get started. We realized that if we wanted to make a significant impact, we would have to narrow our options. So we made two important decisions. One, we confined our efforts to Champion communities where we hoped to stimulate reform at the grass-root level. And two, we focused on middle grade schools."Champion's Middle School

Partnership Program began in Stamford, Connecticut, where the paper manufacturer has its headquarters. The Stamford program implemented the recommendations of the Carnegie report that call for schools to take three major steps: form new and distinct school structures, develop new instructional practices, and foster school-family-community relationships to improve student achievement and aid those at risk of falling behind.

The success of the program in the Stamford Public Schools led Champion to commit itself to establishing similar programs in all of Champion's mill locations.

"This is not a corporate headquarters `top down' program," remarks Eileen McSweeney, the program's current manager, when asked whether Champion employees are involved in the program. "The mill management team must support the partnership and want it in their community." Mill employees serve with teachers, parents, and administrators on the partnership steering committee, which meets regularly to review how the partnership is working in individual schools.

Months are spent discussing the concept of Champion assisting a school district in a mill community to restructure its middle schools so each partner knows what is expected. Champion provides each partnership with a regional director, an educational consultant who works with the principal and teachers on a regular basis to develop the restructuring plan. The plan is based on the Carnegie recommendations, which are adapted to meet the needs of each school in the partnership.

Champion also provides educational consultants to help the schools address particular challenges. "In one of the partnerships," McSweeney explains, "a school decided to stress critical thinking. The partnership brought in a nationally known consultant, Dr. Toni Worsham, to help the school develop a plan to integrate critical thinking throughout the school day. For instance, it's now general practice in this school that when a student is asked a question, there is a 'wait period' which allows everyone in the class to have a chance to reflect on the question and decide on the answer."

In Pensacola, Florida, Champion invited all 600 teachers in the school district to a luncheon to hear firsthand what Champion was proposing and what the partnership would entail.

The teachers responded enthusiastically, expressing a heartfelt "thank you" for the underlying respect this program demonstrates in acknowledging them as professionals with an important role to play. They especially supported the partnership's belief that courses such as Critical Thinking are as vital as the basic curriculum in developing the kind of self-esteem children need to become successful students and adults. According to McSweeney, "Recognizing the importance of teachers and raising the self-esteem of students are critical components of the partnership."

Champion also sponsors annual regional middle school conferences as part of its ongoing support to the partnerships. Nationally known speakers and educators are brought together to conduct intensive staff-development sessions for middle school teachers, administrators, and parents.

The key word used by Jim Hoffman, the program's executive director, when he goes to a mill location to talk to the mill manager, superintendent of schools, and community leaders about setting up a program, is "assist." "We are not there," he says, "to tell schools what to do; we are there to assist them as they work through the Carnegie document."

"Changing attitudes and beliefs is what we are about," McSweeney agrees. "It is a slow process, and we need patience. But we all agree the rewards are extraordinary."

For complete details on Champion's Middle School Partnership Program, see page 124.

HOUGHTON MIFFLIN COMPANY

EDITORS IN THE CLASSROOM

Over the past fifteen years, Houghton Mifflin editors have been going back to school one day a week—not to continue their own education, but to share some of their professional skills with children in twenty-five Boston elementary and secondary classrooms.

It all began when Norma Markson, director of training, was asked to speak at a meeting of professional educators. It was the late seventies, and there was a glut of teachers on the job market. These educators wanted to know how to help get jobs for their students. "After my talk, a professor at Lesley College approached me to discuss an internship program. At that time, Houghton Mifflin, an educational publisher, already had an informal in-house program in place. I wanted to expand it, and in exchange, place our editors in the classroom. From this initial meeting, we developed the exchange program between

Lesley College in Cambridge, Massachusetts, and Houghton Mifflin editors. After the first few years we worked directly with the schools ourselves."

Lesley education students worked at Houghton Mifflin and, in exchange, textbook editors were placed in elementary school classrooms.

Steve Pekich, vice president and director of operations for School Publishing, volunteered for the Editors in the Classroom program in the fall of 1991 to get firsthand experience on how materials are used in the classroom. Pekich was sent to the Eliot School, the oldest elementary school in the North End of Boston. "I had asked to be placed in a first grade class," Pekich explains, "because the first grade is a crucial beginning for a child, and Houghton Mifflin is a major publisher of elementary reading programs." Expecting the role of observer, Pekich was thrown into active participation in the class from the first day. "I floated around the room helping kids with whatever they were doing—matching letters, sounds and pictures, handwriting, reading. I was surprised by the quick and positive manner in which the teacher and her students assimilated me into their classroom environment."

Although the program is aimed at the editors' personal and professional growth, informal yearly self-evaluations show that the teachers benefit from the program as well. Teachers look forward to the Houghton Mifflin editors coming each year. "With the school budget cuts," remarks Pekich, "teachers no longer have teaching aides in the classroom, and our help is greatly appreciated." Teachers also enjoy discussing their ideas for new curriculum and teaching approaches with Houghton Mifflin editors.

"First grade elementary teachers, in particular," recalls Markson, "seem interested in publishing because they publish what their students write. When I worked in the Pierce School in Brookline, Massachusetts, the first grade class was asked to write and illustrate a story. When the story was finished and illustrated and a cover made, it was kept on a special bookshelf of published works."

Houghton Mifflin sponsors an annual reception for teachers. This year it will feature, for the first time, a program on the Evolution of Reading Programs and current trends in integrated language arts. Both teachers and editors benefit by the exchange.

Pekich and Markson have had wonderful experiences in the program and look forward to many more. As Markson exclaims, "I love this program! I think sometimes of going back. If I ever retire, I'd like to volunteer at the Pierce school again."

For complete details on Houghton Mifflin's Editors in the Classroom Program, see page 149.

INGRAM DISTRIBUTION GROUP INC.

INGRAM DISTRIBUTION GROUP/LA VERGNE HIGH SCHOOL LEADERSHIP PROGRAM

Everybody realizes the need to develop leaders: young people with self-esteem and pride who understand the concepts of accountability, planning, teamwork, cooperation, and all the requisites of leadership.

But where do you begin? Where do potential leaders first start to emerge? Ingram Distribution Group Inc. learned firsthand through its involvement with La Vergne (Tennessee) High School. Ingram already had an established Adopt-A-School Program with La Vergne when a teacher, Brenda Royal, asked for help in developing a leadership training program for newly elected student leaders. A meeting was arranged for Royal to meet with Ingram Chairman Philip Pfeffer and Ingram management. They loved the idea and began planning for a leadership conference for the fall of 1991. Forty students attended and heard presentations by Pfeffer and other Ingram senior managers on leadership skills such as communication, time management, and goal setting.

Conference evaluations by participating students, teachers, administrators, and Ingram associates led to the creation of the Leadership Project Program. Students who attended the conference had to come up with original ideas for school improvement projects, take the ideas from inception to completion—just as they would have to do in the real world— and present their completed projects at the 1993 Leadership Idea Fair.

Ingram volunteers chose to sponsor projects on a first-come, first-served basis. Paul Clere, assistant to the chairman, and one volunteer meet regularly with Mark Davis, a senior who conceived the idea of the Campus Beautification Program. "When he first came to me, his plans were grandiose," Clere recalls. "We worked together to break down the elements to see what could be accomplished in the allotted time frame. I'm there to see the project stays on track."

Clere said he and other Ingram volunteers act primarily as consultants. Debbie Webb, accounts payable manager and sponsor of the Disabled Awareness Program, notes her responsibilities have been minimal because the student leading the program, Daphne Davis, has been extremely enthusiastic and innovative. "I met her at a volunteer lunch," Webb said, "and then just talked with her over the phone when she needes a sounding board for ideas."

Davis' primary goal is to raise student awareness of the special challenges disabled persons face. She has designed and taught handicapped awareness lessons to hundreds of students and brought in Bart Dodson, America's Disabled Athlete and winner of eight gold medals in Barcelona, who spoke to a packed auditorium of students for two consecutive class periods. She also challenged drafting classes to design a home to ADA specifications and put the design on display for students to see.

Increasing awareness of special student needs is also a goal of the Teen Pregnancy Support Group Program. Designed to support teens who are pregnant or have children, the program features weekly one-hour group counseling sessions. Speakers are also brought in Wednesdays after school to address topics such as self-esteem, the importance of education, prenatal and infant care, adoption, parenting skills, legal issues, and money management.

Another part of the program is a volunteer student buddy system. Volunteers call their buddies if they miss school and generally provide support and counsel. Nikki Allen, R.N., manager of the Ingram Wellness Center and program sponsor, says the student leaders hope to eventually address pregnancy prevention as well, but that meeting the need for immediate support was considered most critical.

Ingram volunteers like Clere, Webb, and Allen say they get much more out of participating in the L.H.S. Leadership Project Program than they put into it. "The Campus Beautification Project has been a delightful part of my life," Clere said. "I remember the people who made a difference in my life, and believe that time is one of the greatest gifts you can give to anyone."

Clere noted that while volunteers benefit directly from participating, all Ingram associates benefit in some way from the program. "When students not only enjoy but are responsible for the programs and services they need and want, the result is improved morale and school spirit which I think leads to a higher quality education." Clere added that while the entire community benefits from such results in the long run, immediate benefits are enjoyed by the students, many of whom are Ingram associates' children.

Pfeffer, the original program contact, continues to meet with the principal of La Vergne High School and remains closely involved in all Ingram-supported Adopt-a-School activities. "I think it's hard to ask someone else to get involved if you're not ready to give the time yourself," he points out.

"I was interested in developing the leadership training program because I recognized that education and business face the same problems—both students and businesspeople often lack the skills to provide the leadership their organizations and businesses require of them. This is our opportunity to make a difference for young people coming up today and learn a great deal ourselves in the doing."

For complete details on Ingram Distribution Group/La Vergne High School Leadership Program, see page 75.

TIME WARNER INC.

TIME TO READ

Imagine how exciting it must be for a youngster to spend part of the day in the environment of a high-powered publishing company. That's what's happening to young Stanley M. as a participant in the Time to Read program. For this young man, learning to read has become learning for life.

This one-of-a-kind opportunity began for Stanley when Time Inc.'s former chief executive officer, Dick Munro, decided the publishing giant should get more involved in public education. Time Inc. was already in the forefront of reform with its Time to Read program. The company was particularly concerned about the growing literacy problem and was looking for a way to draw on its vast resources. "We wanted to build a volunteer program," recalls Toni Fay, one of the program's founders. "We asked ourselves, 'What do we have that we can donate other than money?' We surveyed the company and found our school-based circulation program that puts *Time* magazine in classrooms across the country."

Time to Read built on the concept of using magazines in the classroom and evolved into a structured reading program providing tutoring and mentoring. Time to Read was originally targeted for adults and was modified so that it could function equally well in a classroom, a workplace, a prison, or a community center.

The program started with six sites in its first year and grew by word of mouth. "The success of this program really rests with the people. I am most proud that it could have happened in a company as diverse as Time Warner," Fay remarks.

"I had just moved to New York from Washington, D.C.," recounts Pattie Sellers, a *Fortune* associate editor, "and wanted to do volunteer work of some kind." She heard about the Time to Read program and the idea of working with kids appealed to her.

"Quite frankly, it was easy. I just had to commit to about two hours each week, and the kids came to our offices."

Sellers has been a tutor since the program started in 1985. She picks up her student, Stanley, in the cafeteria of the Time Warner Building in Manhattan around three o'clock in the afternoon once a week, and they work together in her office until five or so. "There are weeks when I wonder how I'm going to fit it into my schedule," Sellers admits, "but I always manage."

Sellers and Stanley usually check the Associated Press wire on her computer to see the news of the day, read from one of the Time Inc. magazines and complete a Time to Read exercise. They also talk about Sellers's work. "One week I was laying out a six-page story with lots of pictures and charts", recalls Sellers, "and Stanley was able to see how the magazine is put together." The program, in Sellers's opinion, gives kids a new perspective on what work is really like.

A short time ago when Stanley was in Sellers's office, she had to compose her biography and work history for a speaking engagement, and Stanley helped her write it. The next week

Stanley came back and said he wanted to be a writer and that he would start working on his autobiography as his first effort. Now, each week Stanley writes a sentence or two of his autobiography on Sellers's computer.

"If you like kids and feel you should do something outside of work and your social life, this program is perfect for you," says Sellers. "I was very lucky to receive a great education, and now I feel I'm making a real difference in someone else's life."

For complete details on Time Warner's Time to Read Program, see page 169.

EXEMPLARY PROGRAM LISTINGS

Adopt-A-School

1

ACTIVITIES:
Tutoring, donations, guest speakers, special awards.

Program:
Adopt-a-School Program

Program objectives:
To promote reading as fun and encourage both serious and recreational reading at home and in the classroom.

Target group:
Elementary

Year established:
1991

Program description:
In cooperation with the Mayor's Office of Education Services and the Reading Is Recreation program, Bantam Doubleday Dell (BDD) adopted a first grade class at one elementary school in each of the five boroughs of New York. The schools selected were among those with the lowest English and math levels in the city.

Each school is assigned a "captain" from the publisher's staff who works with the respective principals to tailor a plan to meet the needs of the individual schools and to draw on the resources provided by the publisher. BDD donates 1,000 children's books and 500 books for parents to each of the schools, hosts children's author visits, sponsors special events and awards programs such as the Reading Olympics, and provides in-service training seminars for the teachers.

About 100 volunteers from Bantam's three publishing divisions and the corporate office participate in the program. Volunteers are permitted to take time off from work and are provided with transportation to and from the schools.

Publications:
Kidz Power

Annual funding level:
Cost of books and staff time.

Company/organization:
Bantam Doubleday Dell
1540 Broadway
New York, NY 10036

Contact name:
Isabel Geffner

Title:
Vice President and Associate Publisher Dell/Delacourte

Telephone:
(212) 782-8605
See page 31 for success story

2

ACTIVITIES:

Tutoring, incentives, career days, special events, plant tours, student employment.

Program:

Mobil Beaumont Refinery School/Business Partnership

Program objectives:

Improve academic achievement.

Year established:

1989

Target group:

High school

Program description:

Mobil Oil Corporation's Beaumont Refinery, the largest worldwide manufacturing facility, participates in a variety of activities at Westbrooks Senior, the largest high school in Beaumont, Texas. Mobil supports a tutoring program, offers academic and attendance incentive packages, participates in career days, sponsors plant tours, funds special project and special events, and provides employment through the city of Beaumont for incoming seniors and scholarships to students entering the engineering field.

For example, employees are granted one to two hours of release time a week to go to the high school to tutor. Between forty and eighty-five employees participate in the program, tutoring for approximately 1,000 hours. Tutors also initiate special projects such as one refurbishing a Model T car once owned by the man who discovered the first gushing oil well in southeast Texas. Local industrial classes participated in this project with Mobil trade employees.

Mobil also established the Pegasus Enhancement Grant program for teachers. The Pegasus grants provide funds to teachers to supplement classroom instruction and can include requests for materials, training, or equipment. Mobil's partnership activities are coordinated by a full-time employee who works cooperatively with Mobil's Education Committee and the high school principal to review school needs.

Annual funding level:
$10,000

Company/organization:
Mobil Oil Corporation
P.O. Box 900
Dallas, TX 75221

Contact name:
Jeanne Mitchell

Title:
Community Affairs Advisor

Telephone:
(409) 833-9411

3

ACTIVITIES:

Tutoring, special events.

Program:

Newbern Elementary/Junior High Adopt-a-School

Program objectives:

Expand learning opportunities for children and youth and involve employees in schools in the community where they live and work.

Year established:

1989

Target group:

Kindergarten-eighth grade

Program description:

Penguin USA Distribution has been involved in the Newbern (Tennessee) Elementary/Junior High School Adopt-a-School program since 1989. The program is part of the countywide Adopt-a-School program whereby each of the twelve schools in the county is adopted by two or more sponsors. The goal of the program is to improve learning opportunities for children and youth and has the support of all the industries in the area.

Each year Penguin announces the formation of an employee activity committee for the Adopt-a-School program. These volunteers and a school group meet to review the calendar of school events and what the schools need. Penguin employees assist the teachers in the library and classrooms on a weekly basis as tutors, provide judges for science fairs and poster contests, referee intramural games, and participate in a number of other special programs. Penguin presents $50 awards to the most improved student in each of the sixth, seventh and eighth grades each year. In turn the school presents a skit at Penguin's facility, and the volunteers are invited to a Thanksgiving dinner at the school.

There is an Adopt-a-School coordinator at Penguin as well as a paid full-time coordinator for the countywide program. Approximately twelve Penguin employees participate in the program.

Penguin grants employees release time for about half of the activities; the other activities occur after work hours.

Annual funding level:
$2,500

Awards:
Recognized in 1991 by NAPE for excellence in programming and community accomplishment.

Company/organization:
Penguin USA
101 Fabrite Rd.
Newbern, TN 38059

Contact name:
Cindy Rose

Title:
Assistant Benefits Manager

Telephone:
(901) 627-2521

4

ACTIVITIES:
Tutoring, mentoring, career days, plant tours, donations.

Program:
Partners in Excellence

Program objectives:
Improved academic performance and test scores, drop-out prevention.

Target group:
High school

Year established:
1986

Program description:
Proctor & Gamble has been actively involved in the Partners in Excellence (PIE) program, the business/school partnership with the Dougherty County, Georgia, school system, since 1986. Proctor & Gamble's involvement grew out of its recognition that students were falling behind in school, particularly in math and science and not keeping up with trends in technology, which are areas of concern for Proctor & Gamble.

PIE is linked to a five-year strategic plan developed by the Dougherty County School System with specific districtwide goals: increasing the percentage of students completing high school, strengthening staff competence, and expanding learning opportunities to meet the changing demands of society. The program includes tutorial and homework assistance, staff development, and programs in technology.

Fifty to seventy active volunteers from the Proctor & Gamble plant are involved in PIE; volunteer after-work hours range from 150 to 225 hours per week. Plant personnel tutor and mentor students, teach courses and seminars, judge contests, participate in career days, host staff development sessions for the entire school staff, sit on steering committees, and host plant tours. Proctor & Gamble, through its corporate volunteer grant program, also provides funds for computers, software, training materials, and incentives to PIE. The program is coordinated through the Public Affairs Department.

Annual funding level:
$2,500

Company/organization:
Proctor & Gamble Paper Products Company
P.O. Box 1747U.S. Hwy. 19
SouthAlbany, GA 31703

Contact name:
Susan Bryant

Title:
Organizational Effectiveness Group Manager

Telephone:
(912) 883-2000

5

ACTIVITIES:

Tours, guest speakers, demonstrators, donations, special awards, scholarships, career days.

Program:

Partnership in Education, Toyota, T.A.B.C./Jordan High School

Program objectives:

Provide quality education for all students and make a contribution to the community.

Target group:

High school

Year established:

1984

Program description:

Toyota's corporate mission is to get involved in the communities in which the company operates plants. Toyota, T.A.B.C. saw that the best way it could have an impact on the community was through the educational system and formed a partnership with David Jordan High School in Long Beach, CA. Toyota supports a wide range of activities through contributions and staff support.

Toyota annually invites the auto shop classes to tour its plant, which specializes in making truck beds, and introduces students and teachers to the newest technologies. T.A.B.C. gives the auto classes trucks for practice work as mechanics, sends employees into the classroom to demonstrate repair and service procedures, and participates in career days. Students are encouraged to apply for jobs upon graduation and are introduced to the job application process.

Toyota sponsors an annual golf tournament and donates a new truck as a prize, provides special awards and funds for scholarships and special events. The partnership is managed by the Human Relations Department and involves up to five employees.

Annual funding level:
$10,000

Awards:
1992 McKee Award Winner

Company/organization:
Toyota, T.A.B.C., Inc.
P.O. Box 2140
Long Beach, CA 90801

Contact name:
Michael Haruki

Title:
Human Relations Manager

Telephone:
(310) 984-3397

6

ACTIVITIES:

Tutoring, mentoring, in-kind donations, scholarships.

Program:

Tenneco-Jefferson Davis Educational Partnership

Program objectives:

Reduce dropouts, improve academic performance, overcome social skills deficiencies, and raise the number of student planning to attend college.

Target group:

At-risk high school students

Year established:

1981

Program description:

Tenneco's collaborative with Jefferson Davis High School is a three-part, eight-year commitment to more than 1,400 at-risk students and almost eighty teachers. The company plays several roles: volunteer source, in-kind and direct funding agent, and program manager/catalyst. The collaborative pulls together into a focused effort many civic, business and church organizations, Davis's alumni association, students, parents, teachers, and Tenneco employees.

The collaborative is made up of three programs designed to help at-risk students make the transition from middle school through high school to college. The Bridge Program offers students a four-week summer session that links middle school to high school. The Jesse H. Jones Academic Institute, offered in the summers between academic years, introduces students to the principles of leadership, goal setting, and problem solving. Tenneco pays students who attend the institute a stipend and offers them summer employment. And student's entering Jefferson Davis are encouraged to participate in Tenneco's

Presidential Scholarship program, which guarantees graduates four-year scholarships for college if they meet certain criteria.

In 1992 over seventy-five employees contributed more than 7,200 hours of their time to mentor and tutor students. The company provided about $7,000 worth of in-kind services such as printing and gave direct funding of $15,000 for site-based management and matching grants. The program has a full-time manager.

Annual funding level:
More than $375,000.

Awards:
1991 United Technology Exemplary Program Award, 1991 PIE Journal National Partnership Award.

Company/organization:
Tenneco, Inc.
P.O. Box 25115, 1010 Milam
Houston, TX 77252-2511

Contact name:
Jo Ann Swinney

Title:
Director Community Affairs

Telephone:
(713) 757-3930

7

ACTIVITIES:

Incentives, grants, donations, speakers, physical plant improvement.

Program:

The New England/Mather Elementary School Partnership

Program objectives:

Improve test scores in reading and math, improve school attendance, reach at-risk students, increase parent involvement, and support teachers.

Target group:

Elementary school

Year established:

1988

Program description:

Five years ago the principal of the Mather Elementary School was looking to establish a school/business partnership and approached the New England, an insurance company that had been instrumental in establishing the Boston Plan for Excellence in the mid-1980s. The goals of the New England/Mather Elementary Partnership are to improve the success rate of Mather students, increase parent involvement and support teachers.

Through an annual discretionary grant from the New England, a number of educational offerings are made possible. Scholastic magazines and a subscription to the *Boston Globe* are made available to all grades to help improve reading and math scores. Photographs of students with perfect attendance and pizza parties are regular activities aimed at improving school attendance and self-esteem. The New England provides tickets to sporting and cultural events, books for the library, monies for teacher conferences and training programs, and in-kind donations of printing and equipment and furniture when available.

New England employees are involved in the partnership on a project-by-project basis. Sixty employees, for example, painted out the graffiti on the walls of the school. A Celebrity Readers program brings New England executives into the classroom to read to the students. The partnership program is coordinated by an outside consultant and the Public Affairs Department.

Annual funding level:
$5,000

Company/organization:
The New England Public Affairs Department
501 Boylston St.
Boston, MA 02117

Telephone:
(617) 578-6094

Program headquarters:
School/Business Partnership Services
145 Tremont St.
Boston, MA 02111

Contact name:
Alice Jelin

Title:
Consultant

Telephone:
(617) 654-8282

Career Awareness and Leadership Training

8

ACTIVITY:
Guest speakers

Program:
3M Technical Teams Encouraging Career Horizons (TECH)
Program

Program objectives:
To encourage women to pursue scientific and technical
careers.

Year established:
1978

Target group:
Middle and high school students

Program description:
The TECH program, formerly known as the 3M Visiting
Technical Women program, was created to encourage young
women to pursue science and technical careers. Women
engineers and scientists employed at 3M visit local schools
to provide career information, emphasize courses needed for
technical positions, and serve as successful role models for
those considering technical careers. Approximately 100 visits
are made each year to the seventy junior and senior high
schools in the St. Paul-Minneapolis area. More than 100,000
students have been reached since the inception of the
program.

The TECH program operates under the auspices of the 3M
Technical Forum—the technical/professional society within 3M.
A volunteer technical employee manages the program,
matching the schools and volunteers from a list of previous
participants. Approximately 300 employees participate in the
volunteer pool. The volunteer contacts the school directly and
makes arrangements for the school visit. Specially prepared

videotapes and transparencies are available for the classroom presentations. The volunteers pay their own local travel expenses, which are minimal. 3M sponsors a kick-off meeting each year for the program volunteers. Both the schools and the volunteers evaluate the program annually.

Annual funding level:
$1,000

Company/organization:
3M Center Building,
255-2N-01
St. Paul, MN 55144

Contact name:
Judith E. Parker

Title:
Supervisor, Technical Development

Telephone:
(612) 733-9258

9

ACTIVITIES:
Summer internships, field trips, guest speakers, scholarships, teacher training, mentoring, and job shadowing.

Program:
Academy programs

Program objectives:
Career preparedness

Target group:
High school

Year established:
1982

Program description:
The American Express Company created the first Academy of Finance in 1982 to give high school students the opportunity to learn about and prepare for careers in financial services. It is now one of several model programs designed to bridge the gap between the classroom and the workplace.

The academy model consists of a rigorous supplemental academic curriculum that is augmented and reinforced by practical on-the-job experience through paid summer internships. Academy students learn the fundamentals in their chosen field and are immersed in the workplace environment to apply their knowledge to real life situations. Many interns are later hired to work part- or -full time by their internship companies. Program elements include guest speakers and field trips, scholarships, teacher training, and one-on-one mentoring and job shadowing. An academy program can be developed by one company or several companies in an industry.

American Express created the National Academy Foundation (NAF) in 1990 to forge partnerships with professional fields and industries to adapt the academy concept it pioneered in 1982. The Academy of Finance, Ford Academy of Manufacturing Sciences, Academy of Public Service, and Academy of Travel and Tourism operate programs in 30 locations across the country under the guidance of NAF. NAF develops the instructional material for the academy program with industry input and provides in-service training and testing. It costs between $350,00 to $650,000 over a three- to four-year period to develop an academy program.

More than 300 businesses, professional organizations, universities, government agencies, and foundations are partners with NAF in providing financial support, in-kind services, and summer internships. The Fannie Mae Foundation is about to launch the Fannie Mae Academy of Finance in Los Angeles, Miami, and Atlanta (see page 137).

Funding level:
$1 million over five years.

Awards:
More than 20 national, state, and local awards including commendations from two U.S. presidents and three Secretaries of Education.

Company/organization:
National Academy Foundation
235 Park Ave. South
New York, NY 10003

Contact name:
Dr. John Dow, Jr.

Title:
President

Telephone:
(212) 420-8400

10

ACTIVITIES:
Guest speakers, tours

Program:
Career Awareness in Middle Schools (CAMS)

Program objectives:
Career awareness

Target group:
Eighth grade

Year established:
1989

Program description:
The Career Awareness in Middle Schools (CAMS) program provides opportunities for eighth grade students in career classes at West Middle School in Plymouth, Michigan, to interact with UNISYS Corporation employees in a real-world business environment. The eighth grade was chosen because it is the age when students begin to make decisions about the future.

The program is offered as part of a ten-week career course that includes assessment of employability skills, presentations by UNISYS employees on the corporation, job opportunities, and the application and interview process. These sessions are followed by a tour of the UNISYS facility and interaction with company employees who explain their job responsibilities to the students. The objective is to increase student awareness of the workplace and the importance of their education.

Started as a pilot with West Middle School and UNISYS in 1989, CAMS has expanded to include the four other middle schools in the Plymouth Canton Community School District and now reaches all of the district's eighth graders. Each school is partnered with a local business. Ford Motor Company and

Dow Corning, Inc. are among the companies participating in the program.

About ten UNISYS employees participate in the program on a quarterly basis, with volunteer hours estimated at over 100 hours per year. It took about 300 hours to establish the program but the ongoing time commitment to coordinate the program is ten to fifteen hours.

Annual funding level:
Manpower rather than out-of-pocket expenses.

Company/organization:
UNISYS Corporation
41100 Plymouth Rd.
Plymouth, MI 48170

Contact name:
Ron Mack

Title:
Director of Product Assurance & Support

Telephone:
(313) 451-4212

11

ACTIVITY:
Guest speakers

Program:
Choices

Program objectives:
Career awareness, dropout prevention.

Target group:
Ninth graders

Year established:
1983

Program description:
Choices is an interactive classroom seminar program focusing on self-awareness, time and money management, academic decisions, and career consequences. The goal of the seminar is to show students, through classroom exercises and discussion, the importance of education and how academic decisions, such as what courses to take in high school, will effect career and life choices.

Developed by the US West Education Foundation, the two-hour seminar is offered on two consecutive days by trained volunteer speakers from sponsoring corporations. Sponsoring companies pay US West a one-time licensing fee which gives them the right to present the seminar in a specific geographic area. Presenter and classroom kits are purchased from US West. Choices has reached 2.7 million ninth graders and has more 100 sponsors in forty-seven states.

For example, Southern Bell was looking for an opportunity to do something about the high dropout rate in the south and to involve its employees in the community where they live and work, when it was approached to become the pilot site for Choices in the Raleigh, North Carolina, area. The Southern Bell program started in 1989 in Wake County with twenty volunteers presenting the program to 120 students at six schools. The program size has more than doubled and now includes two other counties, with forty-two volunteers presenting the program to 3,800 students at fifteen schools. Southern Bell gives employees release time to make school presentations and employs a full-time coordinator for the two-month period when the program is most active. Annual costs are approximately $5,000, in addition to the one-time licensing fee of $5,000. Choices is now offered by Southern Bell in Kentucky, Georgia and South Carolina.

Annual funding level:
Minimum licensing fee is $2,000 which covers a population base of 200,000. The license is exclusive in a geographic area for five years. Presenter kits are $150; classroom kits are $30.

Participating company/organization:
Southern Bell
P.O. Box 32000
Raleigh, NC 27622

Contact name:
R. David Lane

Title:
Director, Corporate & External Affairs

Telephone:
(919) 881-7690

Program headquarters:
US West Foundation
720 Olive Way, Suite 1725
Seattle, WA 98101

Contact name:
Marge C. Barnette

Title:
Director—Marketing

12

Telephone:
(206) 343-5200

ACTIVITIES:
Internships, guest speakers, special events.

Program:
The Harrison/BOCES Co-op Program

Program objectives:
Career awareness

Target group:
High school juniors and seniors

Year established:
1984

Program description:
The Harrison Co-op Program offers school and work experiences for high school juniors and seniors with high aptitudes in math and science and an interest in pursuing a career in engineering, business administration, or computer sciences. The partnership between the Harrison Division of General Motors and the Orleans-Niagara (New York) Board of Cooperative Educational Services (BOCES) is a two-year work/study program that pairs students with Harrison professionals.

Students alternately work at Harrison in pre-entry level engineering positions for two-week periods and then attend classes at their high school for two weeks. In school the students take accelerated subjects with an emphasis on science and math. At Harrison, they work as apprentices under the direct supervision of engineers, accountants, and industrial managers. Students from fourteen high schools in two counties in New York State are selected for the program on a competitive basis. Of the 106 students who have graduated from the program all have gone to college; eighty-two percent have pursued careers in engineering, business administration;

and computers; and three have been hired as employees. Currently thirty students participate in the program.

About twenty engineers from Harrison act as supervisors/mentors on a rotating basis. Other employees are involved as guest speakers. The program is managed at Harrison by the Personnel Department. Harrison pays the students salaries. This year the program has expanded to include new industrial partners.

Annual funding level:
$3,200 per student with tuition and program costs paid by the school district.

Awards:
1992 United Technology Exemplary Program Award, 1986 New York State Education Dept. Citation for Excellence, 1986 New York State Council on Vocational Education Award.

Participating company/organization:
Harrison Division200 Upper Mountain Rd.Lockport, NY 14094

Contact name:
Ed Hubbard

Title:
Personnel Manager

Telephone:
(716) 439-2011

Program headquarters:
Orleans-Niagara BOCES3
181 Saunders Settlement Rd.
Sanborn, NY 14132

Contact name:
Art Polychronis

Title:

Coordinator

Telephone:
(716) 625-6811 ext. 217

ACTIVITIES:
Leadership conference/fair/mentoring.

Program:
Ingram Distribution Group/La Vergne High School Leadership Program

Program objectives:
Develop basic leadership skills of student participants; empower participants with the ability and accountability to effect changes in their environment and within themselves; and promote self-esteem, pride, and partnership among students, administration and community.

Target group:
High school

Year established:
1991

Program description:
Ingram Distribution Group, Inc. (IDGI) was approached by its adopted school, La Vergne High School, to help it develop a program of student training that would prepare newly elected student leaders for their roles. The program started as a two-day leadership conference led by Ingram's senior management and has expanded to include a leadership project and a leadership idea fair.

Students are selected to participate in the leadership conference based on the merits of a leadership project idea—a school improvement plan they want to implement, and/or faculty nominations. Students develop their projects under the direction of a faculty sponsor and an IDGI advisor who meet with the project teams monthly to review status reports and to advise and assist. Students are responsible for raising the necessary funding to implement their projects. In the spring, the teams prepare project displays, which are viewed by students, faculty, and the community. The school projects are intended to mirror real-life business experiences.

Annual funding level:
$1,200

Awards:
Tennessee Association of School Boards 1992-93 Outstanding Program.

Company/organization:
Ingram Distribution Group, Inc.
1125 Heil Quaker Blvd.
La Vergne, TN 37086

Contact name:
Philip M. Pfeffer

Title:
Chairman of the Board

Telephone:
(612) 793-5066
See page 40 for success story.

14

ACTIVITIES:
Career day programs, job shadowing, scholarships.

Program:
Project SEARCH

Program objectives:
Enable students to plan present and future needs; foster academic growth, career choices and future needs.

Target group:
Junior high

Year established:
1973

Program description:
Project SEARCH (Seeking Experience: A Real Career Help), a week-long project devoted to career awareness, is jointly sponsored by Oakland Junior High School in Columbia and its community partner, Columbia Regional Hospital (CRH). More than 700 students, twenty-five CRH volunteers, and all the Oakland faculty are involved in the program, which focuses on career exploration, career research, interest inventories, job applications and interviews, and job shadowing.

During the week-long program, CRH employees make career presentations in all the seventh and eighth grade science, English, and business classes. Ninth grade classes develop interest inventories, do career research, and complete job application forms. Later in the week CRH employees spend a day conducting one-on-one mock job interviews with each of the 240 ninth graders. The highlight of Project SEARCH is job shadowing whereby students, using their own initiative, arrange a day observing, questioning, and learning about prospective careers. A mentor program has recently been added whereby five students work on a weekly basis in the hospital.

CRH provides about a dozen employees for the full day of job shadowing at the hospital. More than fifteen CRH employees conduct the mock interviews at the school, logging in more than 150 volunteer hours. In addition, two $750 college scholarships are awarded by a joint CRH/Oakland committee. The program is coordinated through the Public Relations Office of the hospital.

Annual funding level:
Minimal

Awards:
1990 NAPE Exemplary Program Award.

Company/organization:
Columbia Regional Hospital
404 Keene St.
Columbia, MO 65201

Contact name:
Beth Morell

Title:
Director of Public Relations

Telephone:
(314) 875-95003

15

ACTIVITY:
Guest speakers

Program:
Youth Motivation Program

Program objectives:
Dropout prevention

Target group:
High school

Year established:
1966

Program description:
The Youth Motivation Program is sponsored by the Chicagoland Chamber of Commerce. The goal of the program is to obtain speakers from local businesses who encourage kids to stay in school. Scott Foresman (SFN) has been participating in the program for ten years.

SFN employees attend an orientation session sponsored by the Chamber of Commerce on how to make presentations to students. Volunteers are asked to speak at a high school during morning hours at least once a year. The schools selected are those with a large minority population and high dropout rates.

Eighteen SFN employees were assigned to one Chicago area school, and each spoke about his or her work and life experience to about three classrooms of students. Students are exposed to the world of work and learn about job opportunities at SFN.

The program is managed at SFN by the Human Resources department who solicits volunteers each year.

Annual funding level:
Covers employees' travel expenses.

Company/organization:
Scott Foresman
1900 E. Lake Ave.
Glenview, IL 60025

Contact name:
Stuart C. Cohn

Title:
Vice President, Human Resources

Telephone:
(708) 657-3986

16

ACTIVITIES:
Job training, field trips, scholarships.

Program:
Career Awareness Program (CAP)

Program objectives:
Lower rate of unemployment among teenagers, provide skills needed for entry level employee.

Target group:
High school seniors

Year established:
1975

Program description:
The Home Savings Bank of America (HSOA) created the Career Awareness Program (CAP) in response to the large numbers of high school seniors who were unprepared for the opportunities and demands of the business world. The program builds a bridge between school and the world of work. The program consists of three phases: classroom training, scholarship, and employment.

CAP classes meet at the school for fifteen weekly two-hour sessions which focus on developing transitional skills and job-preparedness techniques that make students attractive to employees. Field trips to local HSOA branches give students the opportunity to participate in a mock-interview workshop. Approximately twenty-five to thirty seniors from each high school are selected, based on academic achievement, to participate in the program.

Each year one student from each participating high school receives the $4,000 Home Savings of America Scholarship. The recipient is awarded $1,000 per year as long as a 2.5 average is maintained and twenty-four credits of college credit are completed. Home Savings offers summer jobs to the top ten percent of the CAP students from each class. After the summer program, CAP trainees are evaluated. Approximately ten percent of the graduates of the program are hired by HSOA.

One hundred thirty-six schools in California, Florida, Texas, Illinois, and New York participate in the program. Now in its eighteenth year, the program has enrolled more that 15,000 students. The CAP Department of twenty-one persons manages all phases of the program. The department assists other companies that want to implement a CAP at their location. Small businesses can put into effect a portion of the Career Awareness Program, such as summer employment or scholarships.

Publications:
How to Start Your Own CAP.

Annual funding level:
$2 million

Awards:
1989 Partnerships in Education Journal (PIE) Award, 1989 Florida Education Benefactors Award, two national Bellringer awards for Best Community Relations Program.

Company/organization:
Home Savings of America
4900 River Grade Rd.
Irwindale, CA 91706

Contact name:
B. Judy Morgan Phillips

Title:
Vice President, Manager CAP

Telephone:
(818) 814-7211

17

ACTIVITY: MENTORING

Program:
Career Beginnings

Program objectives:
Dropout prevention/job readiness

Year established:
1986

Target group:
High school students

Program description:
Career Beginnings aims to increase the number of high school students from low income families who complete high school and enter college, technical training, or full-time employment.

The program consists of mentoring, summer internships, and special life-skill workshops and is sponsored jointly by business, school, and university partners. Since its inception, more than ninety-five percent of the 14,000 participating students have graduated from high school and well over seventy percent went to college.

Career Beginnings matches each student with a mentor from the local business or professional community. Mentors help students with career planning, applying to college, and understanding the professional work environment. In the summer between junior and senior year, students are offered paid summer internships. In addition, the students attend workshops on career planning, basic academic skills, time and money management, college financial planning, job behavior, and resume writing. The programs generally have a full-time coordinator and are administered by the college or university partner. Career Beginners are mentored by volunteers from dozens of Fortune 500 companies and local corporations such as AT&T, Allen-Bradley Company, Chevron, the Tennessee Valley Authority, Traveler's Life Insurance, and Ohio Bell.

At Thom McAn, for example, employees are invited to a Career Beginnings Fair where they are introduced to the program and have the chance to sign up. Mentors make a two-year commitment to stay with their students from the spring of the junior year through post graduation. Mentors meet with the students at least once a month, at a flexible location, for a minimum of three hours and call several times a month. Thirty-five Thom McAn employees are currently serving as mentors. The company also sponsors twenty-five summer internships, offering students work in stores and in the corporate headquarters. Thom McAn also serves on the Career Beginnings Advisory Committee. The program was started with $50,000 in seed money. Annual costs range from $10,000 to $15,000 plus $25,000 for the summer intern program. Career Beginnings is administered through the Human Resources Department.

Annual funding level:
$3,000 per student plus in-kind donations and services.

Awards:
Career Beginnings is recognized by the American Association of State Colleges and Universities, the California Community Colleges, the Children's Defense Fund. the Conference Board, and the Southern Association of Colleges and Schools, among others, as an exemplary partnership.

Participating company/organization
Thom McAn
67 Milbrook St.
Wooster, MA 01606

Contact name:
Robert Weaver

Title:
Vice President, Human Resources

Telephone:
(508) 791-3811

Program headquarters:
Career Beginnings Center for Corporate
and Education Initiatives
P.O. Box 9110
Waltham, MA 02254

Contact name:
William Bloomfield

Title:
Director

Telephone:
(617) 736-4990

Job Readiness

18

ACTIVITIES:
Workshops, employment, donations.

Program:
Mayor's Summer Jobs Program

Program objectives:
Job readiness

Target group:
High school

Year established:
1967

Program description:
The Summer Jobs Program in Oakland is a twenty-five-year-old partnership between the public and private sector. The program sponsors job-readiness workshops in the public high schools for students and works with local employers to develop summer jobs for youths. The goal of the program is to help students prepare for and find their first job. Matthew Bender has been an active program supporter since 1986, when it moved its business to Oakland.

Volunteers from area businesses participate in the job readiness workshops as corporate trainers. The volunteers go into high school classrooms to teach students how to fill out job applications, prepare resumes, and conduct interviews. The applications are screened before they are sent to prospective employers.

About ten Matthew Bender employees participate in the job readiness workshops as trainers. Matthew Bender regularly hires five to ten youths each summer for clerical positions at salaries of $6 to $6.50 per hour. The Human Resources Department reviews the student applications and coordinates the program. Since 1986, two people have been hired at Matthew Bender as full-time employees. Matthew Bender's in-kind donations include design, art, copy, mailing, and printing services for promotional materials and the annual report.

Annual funding level:
$10,000

Company/organization:
Matthew Bender
2101 Webster St.
Oakland, CA 94612

Contact name:
Diana Pascual

Title:
Director of Human Resources

Telephone:
(510) 446-7217

19

ACTIVITIES:
Tutoring, Guest Speakers, Job Shadowing, Field Trips

Program:
Project STEP

Program Objectives:
To better prepare kids for the transition from school to work.

Year Established:
1971

Target group:
High school

Program Description:
Project STEP (Skills Training Educational Program) provides entry-level job skills training to more than 3,000 California students a year who want to pursue careers in banking. One of the earliest programs of its kind, Project STEP was developed to address the need for a better trained work force. The Bank hires approximately 20 percent of the students trained in the program.

Bank of America develops the curriculum that is relevant to its own operations, provides the training sites and state-of-the art equipment at its bank locations, and the teachers. Project STEP offers over ninety classes in a wide variety of banking and automation skills each school year. Trained bank employees teach subjects such as bank telling, data entry, credit operations, and personal economics. Students sign up for the courses of interest to them and receive a full semester of school credit for each course. The courses run for fifteen weeks, six hours a week. Employees teach classes twice a week after regular work hours and are paid through the school district's Regional Occupational Program. Approximately eighty employees participate in the program as teachers or substitutes and there is a waiting list for new openings. Bank of America sponsors an annual faculty event and presents employee recognition awards. The program is managed through the Department of Human Resources and occupies a third of a full-time employee's time.

Annual funding level:
Under $5,000. Employees are paid by the school district.

Awards:
1986 Presidential Volunteer Action Award for "Best Overall Corporate Effort"

Company/Organization:
Bank of America
333 S. Hope Street
Los Angeles, CA 90071

Contact Name:
Jose Castro

Title:
Vice President and Manager of Education Programs

Telephone:
(213) 345-4324

ACTIVITIES:

Tutoring, guest speakers, job shadowing, field trips.

Program:

Saturday Academy

Program objectives:

Provide academic enrichment to vulnerable twelve- and thirteen-year-olds.

Target group:

Seventh grade

Year established:

1984

Program description:

Saturday Academy is an academic enrichment program for seventh grade designed to help students develop a love of learning and motivate them to continue their education. It is one of the model programs in Aetna's Stepping Up initiative to help disadvantaged young people develop into competent workers. The program is housed in a business or college setting so kids can begin to relate the acquisition of skills to their eventual use on jobs. The program has more than 1,000 graduates.

Saturday Academy runs twice a year for nine or ten Saturdays. Teachers and guidance counselors from area public schools recommend approximately fifty students who are at or near grade level in math and reading. Instruction is provided in math, science, computers, and communications with the focus on integrated hands-on learning. Weekly classes are supplemented by guest speakers, field trips, and job shadowing. Parental involvement is a key feature with parents agreeing to attend half the sessions. Aetna employees volunteer to help out in the classroom, parent workshops, job shadowing, and field trips.

Expenses for meals, trips, materials, and teacher stipends average $30,000 per session.

Developed in 1984 in conjunction with the Hartford (Connecticut) Board of Education, the program has been replicated in Washington, D.C.; Middletown, Connecticut; Atlanta; and Los Angeles.

Annual funding level:
$300,000

Company/organization:
Aetna Life & Casualty Corporation
151 Farmington Ave.
Hartford, CT 06156

Contact name:
Diane Jackson

Title:
Consultant National Issues, Corporate Public Involvement

Telephone:
(203) 273-1932

21

ACTIVITIES:
Guest speakers, job shadowing, field trip.

Program:
Shadow Day

Program objectives:
Job preparation and career awareness.

Target group:
Eleventh grade

Year established:
1991

Program description:
Chevron employees participate in a month-long program with Marietta High School which takes students through the step-by-step process of applying and obtaining a job. The program provides students with the real-life experience of the job-hunting process in a nonthreatening manner. Shadow Day is sponsored annually by the Partners in Education Program, which unites local businesses and schools in support of a variety of programs in the schools.

The program has three components: application, interviewing, Shadow Day. First Chevron makes presentations in English classes describing Chevron's business and available positions. The students work on resumes and fill out the applications with the teachers. Chevron then reviews the applications and selects applicants for interviews. The interviews are held at the high school.

Students come to the Chevron plant for Shadow Day. After an overview of the plant and manufacturing process and plant tour, the students spend one to two hours with their shadows who explain their job responsibilities. Fourteen employees participated in the program last year. Jobs covered in 1992 ranged from instrument technician and process engineer to computer programmer and secretary. Chevron publicizes the program through its electronic mail system. The Shadow Day coordinator at Chevron spends approximately three days preparing for the program.

Annual funding level:
$1,000

Company/organization:
Chevron Chemical Company
P.O. Box 1000
Marietta, OH 45750

Contact name:
Ivin Rohrer

Title:
Training Coordinator

Telephone:
(614) 374-0284

22

ACTIVITIES:
Tutoring, mentoring, curriculum enhancement.

Program:
World of Work

Program objectives:
Help youngsters better understand the relationship between what they are learning in school and what goes on in the workplace.

Target group:
Kindergarten to twelfth grade

Year established:
1990

Program description:
Chrysler Corporation's World of Work program focuses on preparing today's school children for tomorrow's workplace. The program links Chrysler employees with specific public schools located in Chrysler plant cities. More than 2,000 Chrysler volunteers work with students in their classrooms to show the relationship between what they learn in school and how that material can be applied to the work place. The program started at one Detroit elementary school and has expanded to eleven additional schools.

Under the World of Work program, employees are released from work to dedicate an hour or more a week to helping elementary through high school students. Volunteers work with schools to enhance curriculum and serve as role models, tutors and mentors. Volunteers work with students on personal responsibility, attendance, job completion, basic skills, and self-esteem. Training is provided for volunteers and participating schools staff. Each World of Work program reflects the specific needs of the individual school and has a program coordinator.

Annual funding level:

Program is human resource intensive. There is no capital outlay except for the volunteer time and training materials.

Awards:

1990 Point of Light

Company/organization:

Chrysler Corporation
1200 Chrysler Dr.
Highland Park, MI 48288-1919

Contact name:

Valerie A. Becker

Title:

National Education Program Administrator

Telephone:

(313) 956-0607

Mentoring

23

ACTIVITY:
Mentoring

Program:
Adopt-A-Student

Program objectives:
Dropout prevention, increased self-esteem, and improved academic performance.

Target group:
Middle and high school

Year established:
1986

Program description:
The Springfield Institution for Savings (SIS) initiated the Adopt-A-Student Program with Chestnut Middle School to help students at risk of dropping out before completion of middle school. Mentors and students meet for one class period, (forty-five minutes to one hour) each week, at the same time and place at the school. Both students and mentors make a three-year commitment to the program. Teachers and school counsellors select twelve new students each year, and parental permission is required. The curriculum is established by the school coordinator and team teachers based on basic skills and individual needs. SIS hosts special parties and field trips, and mentors often take students out on their own time. Currently thirty-six mentors and students participate in the program.

The mentoring program continues in a less structured form through high school. Students who have completed ninth grade are eligible for summer employment at the bank and after-school jobs during the academic year. The trainee positions are designed to help bring the students up to a point where they qualify for entry-level positions at the bank or in other industries. The bank assists those who choose to apply to college.

SIS employees are carefully screened and go through an orientation program before joining the program. Workshops for the mentors are held during the year on special topics such as self-esteem, English as a second language, and Hispanic culture. The SIS Adopt-A-Student program is managed by a full-time coordinator who also has responsibility for five other smaller partnership programs.

Annual funding level:
$13,000, primarily for student salaries rather than program expense.

Awards:
1991 United Technology Award, 1990 Massachusetts Board of Education Exemplary Partnership Award, NAPE state and national exemplary program awards.

Company/organization:
Springfield Institution for Savings
P.O. Box 3034
Springfield, MA 01102-3034

Contact name:
Denise Laprade

Title:
Partnership Coordinator

Telephone:
(413) 748-8291

24

ACTIVITY:
Mentoring

Program:
Everybody Wins

Program objectives:
Instill love of learning.

Year established:
1988

Target group:
Elementary school

Program description:
Everybody Wins, a New York City mentoring program originally underwritten by the textile industry, was initiated several years ago on a small scale at P.S. 116 in Manhattan. It was based largely on the premise espoused by Jim Trelease (author of *The New Read Aloud Handbook*) and others, that if good literature is read to children, they will perceive that reading is enjoyable and interesting and will become motivated to read. The children selected by the school for the program are generally those who are falling behind in reading and need more personal attention. The program has expanded from fifty volunteers in one school to 200 volunteers from six corporations working in four schools.

Corporate volunteers have lunch in school and read a book with a particular child for one hour once a week. The volunteer works one-on-one with the same child each week. Mentors work with the children in the school cafeteria or in the children's classrooms and select the books to read together. A writing component was added to encourage the students to write to their mentors during the school year and over the summer, Volunteers write to the children if they have to miss a week or when they travel on business or on vacation.

Twenty McGraw-Hill/Macmillan School Division volunteers work at P.S. 59, a school within walking distance of their office. In its second year, the program was introduced on the suggestion of an interested employee. A memo was sent to all School Division personnel describing the program and asking for volunteers. Those interested filled out applications and attended an orientation at the school with the principal and teachers. Everybody Wins works with the school administration to match the students and mentors and establish the volunteer schedule. The program at McGraw-Hill is coordinated by one of the mentors, who is responsible for notifying the school if a volunteer is going to be absent.

Annual funding level:
Administrative costs are $200 per student paid by the corporate sponsors.

Participating company/organization:
Macmillan/McGraw-Hill School Publishing Company
10 Union Square East
New York, NY 10003

Contact name:
Lolita Chandler

Title:
Vice President, National Accounts

Telephone:
(212) 353-5489

Program headquarters:
Everybody Wins Foundation, Inc.
10 Park Ave., Suite 20G
New York, NY 10016

Contact name:
  Arthur Tannenbaum

Title:
  President

Telephone:
  (212) 679-4063

25

ACTIVITY:
Mentoring.

Program:
HOSTS (Help One Student to Succeed)

Program objectives:
Dropout prevention, higher test scores.

Target group:
At-risk students from kindergarten to twelfth grade

Year established:
1972

Program description:
HOSTS (Help One Student to Succeed) is a structured mentoring program in language arts that helps students in kindergarten though twelfth grade with reading, writing, vocabulary development, study skills, and higher-order thinking. Each student is matched with a trained mentor who provides attention, motivation, and support. The mentors are given carefully designed individualized lesson plans drawn from a comprehensive computerized data base. These lesson plans are tailored to a student's learning style, reading level, and motivational interests. Simple instructions are supplied for the mentors and students to guide them in learning activities geared to real-life application.

Businesses provide mentors to existing programs or stimulate interest in the program in the local communities where they operate. For example, several employees at Monsanto's Greenwood, South Carolina, plant had been participating in a local tutoring program that they felt fell short of the needs of students who continued to fail. On their own initiative they researched mentoring programs, found HOST and generated the community support and funding for the program. Currently about 100 Monsanto employees volunteer one hour of their time a week during the workday. The volunteers work one-on-one with the same student each week.

Since its inception, HOSTS has involved over 150,000 students and 100,000 mentors in over 400 programs nationwide. HOSTS provides the data base software, training and ongoing technical support to each school or district that participates in the program. The software monitors attendance, test scores, attitude, and other factors for measuring success that are required for school districts to qualify for Chapter 1 funding.

Annual funding level:
$30,000 first year, $5,000 annual.

Awards:
U.S. Department of Education Mentoring Model, National Center for Dropout Prevention designation as National Model, Chapter 1 National Validation, and numerous national and state awards.

Company/organization:
HOSTS Corporation
1801 D Street
Vancouver, WA 98663

Contact name:
Jerald L. Wilbur

Title:
President & COO

Telephone:
(206) 260-1995

26

ACTIVITIES:
Mentoring, grants, speakers, donations.

Program:
Science and Mathematics (SAM) Mentoring Program

Program objectives:
Enable more students to complete science projects and qualify for scholarships

Target group:
Ninth to twelfth grades

Year established:
1991

Program description:
Westinghouse Electric Corporation has been a partner in the Science and Mathematics (SAM) program at George Westinghouse High, an all-black high school located in Homewood-Brushton, since it was established by the Pittsburgh Public Schools in 1986. SAM is the most comprehensive science and math program in the Pittsburgh city schools and has spearheaded high-technology education in the City of Pittsburgh. The SAM Mentoring Program was started two years ago when the SAMS steering committee recognized that many students in the science and math program were not finishing their science projects and needed help.

Mentors work with students one-on-one three times a month for one-hour sessions. Meetings are scheduled during the school day and are usually held at the local library or at the high school. Mentors assist students with their research and provide guidance and support as they complete their projects. Westinghouse provides students that complete their projects and graduate with a B average a $4,000 scholarship to help them start college. Westinghouse also sponsors an end-of-the year event for students and mentors, and provides in-kind donations of equipment and supplies. Grants also are awarded for special projects. Thirty-seven students and about twenty Westinghouse employees currently participate in the program.

Westinghouse recruits employees through the human resources and division managers who must authorize release time. A one-year commitment is required from the employee and two half-day training sessions are offered before mentors are matched with students. Westinghouse has hired a minority consulting firm that conducts the training sessions for mentors, matches the students and mentors, and publishes the *Mentor Messenger,* a newsletter designed to publicize the mentoring program. Westinghouse also sponsors focus groups through the year so that the mentors and parents can meet. The program is managed by the Contributions and Community Affairs Department.

Publications:
Mentor Messenger

Annual funding level:
$10,000-$15,000 for consultants.

Company/organization:
Westinghouse Electric Corporation
Westinghouse Building
Gateway Center
Pittsburgh, PA 15222

Contact name:
Nina Lynch

Title:
Program Administrator

Telephone:
(412) 642-3627

27

ACTIVITY:
Mentoring.

Program:
World of Work

Program objectives:
Provide inner city youth with insight into the world of work.

Target group:
High school

Year established:
1991

Program description:
World of Work is a mentoring program designed to introduce high school students to publishing and the world of work through the creation of a publication. Cahners staff from all the major publishing departments—editorial, advertising, production, art, office services, human resources, and accounting are paired with students. Students take on all the responsibilities of creating a publication, which is produced on a Macintosh computer.

Students come to Cahners's offices once a week 1 to 3 P.M. for an eight-week period. Students are from the South Bronx Job Corps, an alternative vocational high school. They range in age from thirteen to twenty-three and are selected for the program based on good attendance and their academic record. The students work together as a group for an hour a week and with their mentor for the other hour. The students receive academic credit for the program. The program is offered once a year. The mentors continue to maintain contact with their students on an informal basis after the program is completed.

The program is offered in conjunction with the New York

City Public Schools Office of External Programs and Career Education Center, which operates alternative education centers.

Publications:
Once in a Lifetime

Annual funding level:
$1,500

Company/organization:
Cahners Publishing Company
245 W. 17th St.
New York, NY 10011

Contact name:
Loriann Weiss

Title:
Recruitment Manager

Telephone:
(212) 463-6624

School Reform
and Curriculum
Enhancement

28

ACTIVITIES:

Mentoring, tutoring, internships, leadership training, curriculum development, staff development, scholarships.

Program:

Kodak 21st Century Learning Challenge

Program objectives:

Improve math and science achievement for all students

Target group:

Preschool to twelfth grade

Year established:

1990

Program description:

The Kodak 21st Learning Challenge is a ten-year partnership between the Eastman Kodak Company and pre-college educational institutions in Kodak plant communities nationwide to improve fundamentally mathematics and science achievement for all students. Kodak's commitment to the Rochester (New York) City School District, the flagship for the Learning Challenge and Kodak's hometown, involves nearly 1,500 Kodak employees working as mentors, tutors, and team-teachers, with more than 5,000 students in classrooms and Kodak facilities.

The Kodak Learning Challenge is made up of six elements: (1) school partnerships (there are currently nineteen of them with Rochester schools), (2) a mentor program, (3) the Summer Science Work-Study Institute for teachers to develop a cadre of math and science specialists, (4) early childhood education centers to create a preschool curriculum, (5) an Implementation of Quality Education Process to help schools design and execute school improvement plans, and (6) volunteer training in teaching techniques for Kodak employees.

The mentoring program is a key component of the Kodak 21st Learning Challenge. The program links student, parent, teacher, and Kodak employee in a six-year learning commitment that extends from fifth/sixth grade until graduation from high school. The program includes monthly visits to the mentor on the job site to introduce students to career opportunities.

Representatives from schools, district offices, the community, and Kodak plan, implement, and monitor the program at each site. The program is being developed in Kingsport, Tennessee; Longview, Texas; and Windsor, Colorado, with plans to implement the program in other plant cities as well. Program headquarters in Rochester has a full-time staff of seven to support local program activities and publish a national newsletter, *Star Chronicles*.

Annual funding level:
$2.5 million. The 21st Learning Challenge was recently awarded a $1.4 million grant from the National Science Foundation for minority student achievement.

Publications:
Star Chronicles

Awards:
1992 National Alliance of Business, Business Education Partnership Program of the Year; 1992 Anderson Merit Award; 1991 Conference Board Best in Class Award.

Company/organization:
Eastman Kodak Company
343 State St.
Rochester, NY 14650-05454

Contact name:
Jodie Belcher

Title:

Manager, Kodak 21st Learning Challenge

Telephone:

(716) 724-2785

Special Awards,
Incentives, and
Scholarships

ACTIVITIES:

Tutoring, mentoring, student employment, incentives.

Program:

Cities in Schools/Burger King Academy

Program objectives:

Dropout prevention

Year established:

1977

Target group:

At-risk kindergarten to twelfth graders, with emphasis on high school.

Program description:

City in Schools, Inc. (CIS) is the largest nonprofit dropout-prevention program. The CIS method brings small teams of repositioned social service providers into schools where they can form one-on-one relationships with students and work alongside teachers, volunteers, and mentors to keep children in school. A training institute was established so that every community that desires a CIS dropout-prevention program can send representatives, including business leaders, to be trained in how to start and operate one.

There are currently sixty-nine operational local programs serving 131 communities in twenty-two states. CIS programs operate at 433 educational sites. The programs are overseen and developed by a board of community, business, and school leaders and are 501(C)3 (not for profit) corporations. Programs can encompass a city or a county or a district. A paid citywide executive director oversees the educational sites within a program and negotiates with the different institutions providing services.

In 1989 Burger King joined Cities in School to form a national network of academies designed for students who have already dropped out of school or who are functioning below their potential in a traditional school. Unlike the traditional CIS model, the Burger King Academies are housed in a separate wing of an existing school or in a separate facility altogether and provide a unique, individualized, and supportive environment. There are currently twenty-seven Burger King Academies in fifteen states and two countries.

The academies are underwritten in the first year through corporate contributions and thereafter supported by franchisee fund-raising efforts. City in Schools develops the program in each market, and Burger King franchisees serve on the local board of directors. Burger King franchisees also serve as mentors, sources of employment, and guest speakers and provide academic-achievement incentives such as free tickets to local events and food coupons and scholarship funds.

The national organization is a public/private partnership supported by a variety of private businesses, foundations, and individuals, as well as an interagency grant from the U.S. Departments of Justice, Labor, Health and Human Services, and Commerce. Corporations provide funding, mentors, and tutors to existing local programs.

Annual funding level:
Budgets vary. Salary of executive director can range from $10,000 to $45,000. The budget for a Burger King Academy is approximately $500,000 per year.

Awards:
Certificate of Special Recognition for national leadership from the Business-Higher Education Forum of the American Council on Education; daily Point of Light for local program.

Participating company/organization:
Burger King
P.O. Box 520783
Miami, FL 33152

Contact name:
Richard Fallon

Title:
Director of Corporate Involvement

Telephone:
(305) 222-8862

Program headquarters:
Cities In Schools, Inc.
401 Wythe St., Suite 200
Alexandria, VA 22314

Contact name:
Cordell Richardson

Title:
Assistant to the Executive Director/Coordinator of Regional
Activities

Telephone:
(703) 519-8999

30

ACTIVITIES:

Teacher training, curriculum design, teacher workshops and conferences.

Program:

Middle School Partnership

Program objectives:

To provide a superior education to every middle school student in the program.

Target group:

Middle school, ages ten to fifteen.

Year established:

1989

Program description:

Champion International, a paper manufacturer, launched a partnership with the Stamford (Connecticut) Public Schools in response to the recommendations in *Turning Points*, the 1989 Carnegie Report on Adolescent Development. This report identifies ages ten to fifteen as the last best chance to reach kids and improve their chances of educational success. Champion's goal was to assist in the restructuring of the middle schools. The initial plan involved the reorganization of the three middle schools from the K-6, 7-8, 9-12 matrix to the K-5, 6-8, 9-12 format and the creation of a magnet middle school to focus on math, science, and technology as well as programs to assist at-risk students.

Now in its third year, the Stamford program has served as a model for Champion programs in middle schools in the Upper Peninsula of Michigan and Pensacola, Florida, where Champion has large paper mills. Champion hosts annual Middle School Partnership Conferences in each location to provide middle school teachers, administrators, and parents with educational programs and workshops led by national experts. There are plans to establish programs in Champion's twelve other mill locations in the future. Champion makes a five-year commitment to each partnership program it establishes and employs three persons on a full-time basis to develop and manage its Middle School Partnership Program.

Annual funding level:
  $1.5 million

Awards:
  1992 Partners in Education (PIE) Award

Company/organization:
  Champion International Corporation
  One Champion Plaza
  Stamford, CT 06921

Contact name:
  F. James Hoffman

Title:
  Executive Director, Middle School Partnership

Telephone:
  (203) 358-2815
  See page 34 for success story.

31

ACTIVITIES:

Career day programs, tutoring, leadership management training, workshops, curriculum design, donations.

Program:

Motorola, Inc. Land Mobil Products Sector Partnership with Illinois School District U. 46

Program objectives:

Academic achievement, increased enrollment in math and science, dropout prevention, teacher development, curriculum reform.

Target group:

Kindergarten to twelfth grade

Year established:

1990

Program description:

The Motorola/School District U. 46 partnership is twofold. On a district level, the focus is on mathematics and science-curricula reform and instruction for kindergarten through twelfth grade. The partnership with Larsen Middle School, the Hispanic bilingual middle school for the district, focuses on all phases of educational development for students, teachers, parents, and staff.

Programs target staff development, students, and parents. For example, Motorola offers teachers summer employment and sponsors a number of special workshops for guidance counselors, teachers, and principals. Motorola engineers are partnered with teachers to implement the Motorola Electronics Kit in high school physics classes and eighth and ninth grade physical science classes. Special programs are also offered at the Motorola Electronics Museum. Family English as a Second

Language classes for which babysitting was offered were started at Larsen, and other programs are being initiated to involve parents in the school. Motorola also donates furniture and equipment to schools.

The partnership is intended to be a replicable model for other businesses and community organizations to use in developing partnerships to support long- and short-term strategies to improve schools. School District U. 46 is the second largest school district in Illinois serving 28,000 students. Several hundred Motorolans participate in various partnership activities, which are coordinated at Motorola by a full-time program manager.

Annual funding level:
Covers salary and expenses of Program Coordinator and systemic projects as needed.

Awards:
1992 Illinois Exemplary Business/Education Partnership Award.

Company/organization:
Motorola, Inc.
1301 E. Algonquin Rd
.Schaumburg, IL 60196-1065

Contact name:
Beth Tinnmons

Title:
Education Relations Manager, Land Mobile Products Sector

Telephone:
(708) 576-3362

32

ACTIVITIES:

Special events, assemblies, guest speakers, demonstrators.

Program:

SC Johnson Wax Kaleidoscope Educational Series

Program objectives:

Enhance and expand children's classroom learning experience

Year established:

1979

Target group:

Kindergarten to eighth grade

Program description:

SC Johnson Wax developed the Kaleidoscope Educational Series to enhance and expand children's classroom learning. The goal of the program is to provide supplemental educational experiences for Racine and Kenosha county (Wisconsin) students through a series of dramatic and unique programs. Children from public and private schools are invited to attend at no cost a variety of curriculum-based programs. Due to the popularity of the program, interested teachers fill out reservation forms and a lottery determines which schools will be able to attend.

The programs combine speakers, demonstrations, visual aids, and audience participation to educate students on specific topics ranging from science and technology to fine arts and environmental education. Many of the programs are focused on subjects and skills required by business. The Kaleidoscope programs for the 1992-93 school year, such as Innovations in Science Technology and Robotics and Close Encounters of the Chemical Kind focus on science, computer technology, and the environment, reflecting three of SC Johnson Wax's corporate objectives.

Since its inception in 1979, more than 350,000 students in kindergarten through eighth grade have attended the programs held at the SC Johnson Wax Golden Rondelle Theater. To measure effectiveness, teachers and chaperons are surveyed and the results tabulated. The program is developed in conjunction with the educational community. Costs for each program, including speakers, vary between $1,000 to $5,000 with some programs, such as one from the National Aeronautics and Space Administration, being free. The program is managed by a part-time employee in the Guest Relations Department, which is part of the Corporate Communications Department.

Annual funding level:
$14,000 for program presenters. Allocated costs for staff time and the theater are not included.

Awards:
1990 Environmental Achievement Award

Company/organization:
SC Johnson Wax
1525 Howe St.
Racine, WI 53403

Contact name:
Kari Iselin

Title:
Community Relations Project Coordinator

Telephone:
(414) 631-2021

Teacher Training

33

ACTIVITIES:

Curriculum design, guest speakers, demonstrators.

Program:

Dow Hands on Science Project

Program objectives:

Improve the teaching of science in the nation's schools.

Target group:

Elementary school

Year established:

1989

Program description:

Spearheaded by Dow's then-senior vice president and chief scientist David Sheetz, Dow Chemical Company has supported the efforts of the National Science Resource Center to develop a hands-on science curriculum for grades one to six since 1987. The first hands-on project was implemented in 1989 with the Midland, Michigan, public schools, where Dow has its headquarters. This project became the model for the thirteen teams that are now in place across the country.

First a leadership team is selected composed of the superintendent of schools, the science coordinator, a dedicated teacher, and a Dow scientist. The team attends a leadership institute at the National Resources Center of the Smithsonian Institution. A five-year plan is drafted at the leadership meeting that becomes the districtwide implementation strategy. Dow works with the team over six years and provides human resources, time, and money. For example, Dow and the Fresno (California) Unified School District (FUSD) have been partners since 1988. The initial target group of 1,000 students in kindergarten to sixth grade in two schools located in low-income areas of the district has expanded to eleven more schools and is expected to cover all sixty FUSD elementary schools by 1996. To support the program, Dow has committed $40,000 per year until 1996.

About 100 Dow employees are working in hands-on science programs in school systems in the U.S. and Canada. Dow has a long-term goal of involving at least one school district at every Dow site.

Annual funding level:
Approximately $250,000 per team for a six-year period.

Company/organization:
The Dow Chemical Company
2020 Dow Center
Midland, MI 48674

Contact name:
Jan B. Loveless

Title:
Manager of Education Affairs

Telephone:
(517) 636-2471

34

ACTIVITY:
Special awards

Program:
Book It!
Program objectives: Motivate children in the elementary grades to read more, both at home and at school, and to help pre-readers develop a positive attitude toward learning to read.

Target group:
Kindergarten to sixth grade

Year established:
1984

Program description:
Book It! is a five-month reading incentive program sponsored by Pizza Hut for students in kindergarten through sixth grade. The program, which reaches about 16 million students, motivates children to read more at home and in school by rewarding them for their reading accomplishments. Over 700,000 classrooms participate in the 1992-93 program, which is supported by all the Pizza Hut franchises across the country. The national program is supported by four staff members.

At the outset, teachers set monthly reading goals. Goals can be set for each student or for a whole class. These goals can be number of books read, number of pages read, number of minutes read, etc. For students who are not yet reading or who have difficulty reading, goals can be set where parents or others read to them. Teachers and parents verify that reading assignments are satisfactorily completed. As soon as the monthly goal is met, the teacher gives the student a Pizza Award certificate which is redeemable at any participating Pizza Hut.

On the first visit, the restaurant manager gives the child a button and a star to place on the button. The child receives another pizza and a star each month the reading goal is met. If a child meets the reading goal for all five months, he or she is placed on the Book It! Reader's Honor Roll. If all children in a class meet their reading goals in any four of the five months of the program, the entire class and teacher are given a free pizza party.

Annual funding level:
$20 million

Awards:
More than twenty national, state, and local awards including commendations from two presidents and three Secretaries of Education.

Company/organization:
Pizza Hut
P.O. Box 2999
Wichita, KS

Contact name:
Eunice Ellis

Title:
Director

Telephone:
(316) 681-9797

35

ACTIVITIES:

Scholarships, mentoring, internships, business visitations, workshops, tours.

Program:

Fannie Mae/Woodson Senior High School Incentive Scholarship Program

Program objectives:

Improve student achievement and increase the number, types, and quality of post-secondary opportunity.

Target group:

High school

Year established:

1988

Program description:

In 1988 the chief executive officer of Fannie Mae made a pledge to the students of H.D. Woodson Senior High School, located in one of the most depressed economic areas in Washington D.C., that the company would provide at least one million dollars, mentors, time, and jobs over the next ten years to ensure their opportunity for higher education. Since that pledge was made, more than 280 students have been inducted into the Incentive Scholarship Program.

When a student earns all As and Bs in a semester, $500 is put in a special account for that student's post-high school education. The maximum award is $4,000 plus accrued interest. That student is inducted into the Future 500 Club, assigned a mentor, and given the opportunity to participate in workshops and cultural events. The Mentor Program is an important component of the success of the program. Mentors (granted ten hours of activity leave), serve as role models and offer support to their proteges. More than 200 Fannie Mae employees participate in the mentoring program. After their freshman year in college, students are also eligible for summer internships at the Fannie Mae, which combines training and development with work experience.

Fannie Mae plans to replicate the program in south central Los Angeles, Miami, and Atlanta using the Academy of Finance model developed by American Express under the direction of the National Academy Foundation (see listing on page 67). The scholarship incentive program is administered by full-time staff who report to the Community Relations Department. Outside consultants are used on a project basis.

Annual funding level:
$265,000 program costs; scholarships are an additional $250,000.

Awards:
D.C. Partners in Education Award

Company/organization:
Fannie Mae
3900 Wisconsin Ave., NW
Washington, DC 20016

Contact name:
Lonnie Edmonson

Title:
Manager, Fannie Mae/Woodson Senior High School Incentive Scholarship Program

Telephone:
(202) 752-7850

36

ACTIVITY:
Scholarships

Program:
I Have A Dream

Program objectives:
To motivate disadvantaged youngsters to stay in school and guarantee to every child the opportunity and funds for higher education.

Target group:
Elementary school

Year established:
1981

Program description:
In 1981, New York city businessman Eugene Lang adopted an entire sixth grade class at P.S. 121 in East Harlem and said he would send them to college if they stayed in school. Lang's initial experience evolved into the I Have a Dream Program (IHAD Program). The IHAD Program gives sustained care, personal support, and guidance to disadvantaged and at-risk elementary school children (Dreamers) from elementary school through high school, plus scholarships support needed to assure a college or vocational opportunity.

First, an individual sponsor or a small group of sponsors adopts an entire inner-city elementary school. Second, the sponsor provides or raises the minimum of $350,000 required to fund the project. Third, in cooperation with local public school authorities, an inner-city fourth, fifth or sixth grade class is selected (forty to sixty students). Fourth, a project coordinator is hired to develop and maintain day-to-day contact with the Dreamers, their families, and the schools. Finally the sponsor arranges for the project to have office and activity space in an accessible community service facility, school, or college. Sponsors develop a program of academic, cultural and social support activities for their Dreamers.

Today more than eighty percent of Lang's original class have their high school or general equivalency diplomas, and Lang's experience at P.S. 121 has become the model for IHAD programs across the country. Today there are almost 200 sponsors of 159 programs in 48 states involving over 10,000 Dreamers. The IHAD Foundation oversees and supports these programs across the country and assists in the creation of new projects.

Annual funding level:
One-time commitment to provide or procure $350,000 to sponsor an entire elementary school class.

Company/organization:
I Have a Dream Foundation
330 Seventh Ave.
New York, NY 10001

Contact name:
Timothy Arena

Title:
Director of Communications

Telephone:
(212) 736-1730

37

ACTIVITY:
Special awards
Program:
Middle School Attendance Incentive Program

Program objectives:
Improve school attendance.

Target group:
Middle school

Year established:
1989

Program description:
The Middle School Attendance Incentive Project was developed by the Washington County, Maryland, Board of Education and McDonald's Family Restaurants to reinforce positive school attendance and to enhance self-respect in students who attend school regularly by granting incentive awards. Students in the sixth, seventh, and eighth grades are encouraged to "go for the gold," silver, or bronze individual awards and for homeroom and school awards.

The program is centered around the concept of the CIA—the Committee to Improve Attendance. Individual, homeroom, teacher, and school goals are established and awards given for increases in overall attendance, in the number of students with perfect attendance, and in the number of students who miss no more than two days per year and for a decrease in the number of absences.

McDonald's provides the incentives and supports the community in its educational goals. Individual awards include credit at McDonald's, homework waivers, and free assignment and qualification in special drawings. Homeroom awards include a "celebration party" at the end of the semester at a local McDonald's Restaurant. Schools can win banners, plaques, and $250 contributions toward educational equipment.

More than 4,400 students from eight middle schools in the county participate in the program. Statistics show that attendance has continued to increase in each of the three years the program has been in place. The program has been used as a model for other school incentive programs in Maryland. This program originated as part of a local McDonald's public relations program. It is one of four education programs implemented by McDonald's on a regional level and managed by an outside public relations firm.

Annual funding level:
Approximately $5,000 (direct costs only).

Awards:
National Association of Pupil Personnel

Company/organization:
McDonald's Family Restaurants
3015 Williams Dr.
Fairfax, VA 22031

Contact name:
Kathy Foster

Title:
Vice President, Director of Public Relations

Telephone:
(703) 698-4000

38

ACTIVITY:
Special awards

Program:
Scholastic Art and Writing Awards

Program objectives:
Encourage creativity and achievement in art and writing.

Target group:
Seventh through twelfth grades

Year established:
1922

Program description:
The Scholastic Art and Writing Awards program is the largest competition of its kind in the nation, drawing more than a quarter million entries each year from all fifty states and abroad. The Scholastic Art and Writing Awards combined offer prizes in more than twenty categories to attract the broadest participation and to encourage young people to explore a wide range of genres. Every student that participates in the awards program receives some form of recognition.

Promotional mailings announcing the program and entry rules go to English and art department in schools across the country. The public schools seek local sponsors to mount regional art competitions of The Scholastic Art Awards, which are locally funded. Winners of these seventy sponsored regional competitions qualify for national judging in the spring. Winners of The Art and Writing Awards are selected by juries of leading artists, writers, educators, and Scholastic editors.

The national finalists are recognized in a national student art exhibit mounted at a leading art institution, such as the National Gallery of Art in Washington, D.C., and the Carnegie

Museum of Art in Pittsburgh. Outstanding works are published in Scholastic magazines.

The program awards more than $90,000 in scholarships and helps winners secure an additional $350,000 in scholarships from more than fifty college and universities. The program has grown to become a nationwide effort, supported by thousands of teachers, parents, and community and business leaders. National sponsors including The Hallmark Corporate Foundation, Smith Corona, Strathmore Paper Company, and *The New York Times* provide in-kind support and underwrite program awards. The Scholastic Art and Writing Awards program is administered by a full-time staff employed by the Scholastic Foundation.

Publications:
> *The Scholastic Art Awards*; SCHOLASTIC MAGAZINES—*Art, Literary Cavalcade.*

Annual funding level:
> Information on national sponsorship costs is available on request.

Company/organization:
> Scholastic, Inc.
> 730 Broadway
> New York, NY 10003

Contact name:
> Susan E. Ebersole

Title:
> Director

Telephone:
> (212) 505-3404

39

ACTIVITY:
Special awards

Program:
Thumbs Up

Program objectives:
Student achievement, higher graduation rate, and increased parent involvement.

Target group:
Special needs students in elementary, middle, and senior high school.

Year established:
1987

Program descriptions:
Thumbs Up is a partnership between Sunbank/Miami N.A. and the Dade County (Florida) Public schools aimed at motivating learning-disabled exceptional children through an incentives program. The program started when an elementary school guidance counsellor went to the Sunbank branch across the street from her school and asked the branch manager for help for her special education students. The program now includes eighty-eight volunteers from all twenty-eight branches of Sunbank and reaches over 2,600 students in fifty-one schools.

Awards are presented for perfect attendance, all passing grades, and achieving the honor roll. Incentives such as Thumbs up appliques, certificates, luncheons, and pizza parties with the students and branch managers and/or Sunbankers are awarded every six weeks at a breakfast hosted by Sunbank volunteers. The culminating event of the year is the Outstanding Student Awards Ceremony which recognizes three students from each of the participating partner schools. Branch managers participate in career day fairs at their respective schools and conduct banking classes on check writing/money management. Branches also display student art projects.

Thumbs Up is coordinated by Sunbank's marketing department. The branches submit a program budget to the corporate marketing department, which purchases the incentives and hosts the Outstanding Awards ceremony. The branch managers maintain direct contact with their local school partners.

Annual funding level:
$10,000

Company/organization:
Sunbank/Miami N.A.
777 Rickell Ave.
Miami, FL 33131

Contact name:
Lucy Rivera

Title:
Marketing Specialist

Telephone:
(305) 579-7198

40

ACTIVITIES:
Workshops, conferences, business visitations.

Program:
Cray Academy
Program objectives:
Improve training of math, science, and technology teachers.

Target group:
Teachers in kindergarten to twelfth grade

Year established:
1988

Program descriptions:
The Cray Academy is a two-week program designed to help teachers of kindergarten through twelfth grade promote literacy in mathematics, science and technology. The academy program, funded by Cray Research, Inc., is part of the Wisconsin Educational Partnership Initiative, which was created to improve the math, science, and technology curriculum and instruction in western Wisconsin public schools.

More than 1,100 teachers participated in the fifth annual academy held in the summer of 1992 in Chippewa Falls, Wisconsin.

Outstanding speakers keynote each week of the academy, where workshops are conducted by leading educators and experts on such topics as hands-on mathematics, science and technology education, integration, cooperative learning, critical and logical thinking, high-tech communications, and new forms of assessment. The academy gives teachers the opportunities to test concepts and share ideas with many other teachers from the region and with experts around the nation. The academy stresses the need for teachers to be exposed to the corporate world. Teachers take tours to businesses in three different industries. Staff development programs and follow-up activities are available to teachers who have attended the academy.

Participants in Cray Academy are selected by their respective school district administrators. Funding is provided by Cray Research and federal monies allocated to each school district. Participants have the option of earning graduate credits or continuing-education hours for licensure for each week of attendance from the University of Wisconsin-Eau Claire.

Annual funding level:
$250,000Awards: Several state awards

Company/organization:
Cray Research, Inc.
655A Lone Oak Drive
Eagan, MN 55121

Contact name:
Bill Linder-Scholer

Title:
Director of Community Affairs

Telephone:
(612) 683-7386
Program headquarters:Cray Academy
1345 Ridgewood Dr.
Chippewa Falls, WI 54729

Contact name:
Julie C. Stafford

Title:
Executive Director

Telephone:
(715) 723-1181

41

ACTIVITY:
Staff/teacher training

Program:
Editors in the Classroom

Target group:
Elementary and secondary school

Year established:
1976

Program objectives:
Professional development of editors.

Program descriptions:
Editors in the Classroom is an in-classroom program designed to give editors classroom experience and the opportunity to interact with teachers and students in local schools. The objective of the program is professional development.

Editors are placed in twenty-five Boston-area elementary and secondary schools based on the editor's editorial interest and grade level. Editors are in the classroom one morning a week and usually make a commitment of one school term (eight to ten weeks). Editors help with lessons or individual students, lead classes, give presentations, and act as teacher's aids. The program is flexible and gives teachers the opportunity to discuss curriculum. Approximately five to fifteen editors participate in the program each year. Teachers volunteer for the program.

Editors in the Classroom started in 1976 as a student exchange program between Houghton Mifflin editorial staff and education students at Lesley College in Cambridge, Massachusetts. Lesley students worked at Houghton Mifflin, and HM editors were placed in local classrooms. The program is now managed by a training editor who arranges the placements with a school administrator or department head; student internships are handled as a separate program. Houghton Mifflin holds a yearly reception for participants where experiences are shared and books are given as gifts.

Annual funding level:
Minimal.

Company/organization:
Houghton Mifflin Company
One Beacon St.
Boston, MA 02108

Contact name:
Norma Markson

Title:
Director of Editorial Training

Telephone:
(617) 725-5387
See page 37 for success story.

42

ACTIVITY:
Teacher training

Program:
Texas Instruments Exchange Program

Program objectives:
Teacher training

Target group:
High school math and science teachers

Year established:
1992

Program descriptions:
The Texas Instruments (TI) Exchange Program, in partnership with the Ector County (Texas) Independent School District, offer secondary teachers with a degree in mathematics, chemistry, physics, or the physical sciences the opportunity to apply for a one-year developmental leave to work as an employee of Texas Instruments. The objective of the program is to give teachers firsthand experience of the workplace that can be taken back to the classroom and shared with other teachers.

The Executive Council of the Ector County School District recommends teacher candidates for consideration based on education, service (minimum of five years is required), ability to perform in a team environment, career ladder status, and the availability of a replacement. TI conducts the next level of screening and makes the final selection. The selected teacher is regarded as a TI employee but does not lose teacher retirement status or other benefits. The selected teacher continues to receive his or her base salary from the school district. The teacher also agrees to return to the school district for a minimum of three years, while TI agrees not to hire an intern for five years.

The internship program was developed as a six-month pilot and is managed by the Human Resources Department. TI provides monthly supervisory training and computer instruction to the teacher, who works in a supervisory capacity as part of a work team. TI is considering extending the internship program to another school district.

Annual funding level:
One-half of teachers salary.

Company/organization:
Texas Instruments
P.O. Box 60448
MS 3013
Midland, TX 79711

Contact name:
Kelvin Michael Alexander

Title:
Human Resources Administrator

Telephone:
(915) 561-6636

Tutoring

43

ACTIVITY:
Tutoring

Program:
Boeing Homework Club

Program objectives:
Improve student achievement in math, reading, and social studies.

Target group:
Elementary and junior high school

Year established:
1989

Program descriptions:
The Boeing Homework Club is an after-school tutoring program for elementary and junior high students in the Auburn School District in Auburn, Washington, who are failing or just passing in school. Boeing employees volunteer their time two days a week, working with one or two students for forty-five minutes. At the elementary level, grades three through six, the tutoring sessions focus upon the students' homework in reading, math, and social studies. At the junior high level, the tutors assist the students with math. More than 500 students and 250 volunteers have been involved in the program since its implementation.

About sixty Boeing employees from the Fabrication Division, ranging from engineers to hourly office workers, currently participate in the program. The Boeing volunteers are actively recruited and trained within the company; the students are carefully selected, referred, and matched with a tutor by qualified district staff. The tutoring sessions are held at the schools, which are located near the Boeing facility. The program was conceived and developed by the Boeing educational representative and the assistant superintendent of elementary education, is now managed by the Community Relations Department.

Annual funding level:
$1,500

Company/organization:
The Boeing Company
P.O. Box 3707 MS 5E-28
Seattle, WA 98124-2207

Contact name:
Bonnie Hogan

Title:
Program Administrator

Telephone:
(206) 931-9064

44

ACTIVITY:
Tutoring

Program:
Children of the Future

Program objectives:
Imbue eagerness for learning; raise math and reading levels.

Target group:
First though tenth grades

Year established:
1990

Program descriptions:
Children of the Future is a community-based tutorial program for at-risk students. Approximately 150 students attend one of four one-hour tutoring sessions held Saturday morning and two evenings a week at North Chicago Community High School. Teachers refer students that are falling behind in reading and math. Parental permission is obtained, and the tutors and students are matched. Volunteers indicate the age level they want to work with. Improvements have been seen in reading and math, self-esteem, completion of class work, motivation, and self-confidence. At the close of each year, a recognition breakfast is held for the students, teachers, parents, volunteers, and the community. Certificates of participation and awards are given to the tutors and students.

Abbott Laboratories was approached in 1989 by community leaders who were concerned about the low levels of ACT scores in county schools. Abbott made an initial $5,000 donation to the start-up effort and provides continuing funding for a program coordinator, materials, and refreshments. Two-thirds of the 160-member volunteer staff are Abbott employees. The military, the city, and other businesses also contribute to the partnership.

Awards:
1992 United Technology Award

Annual funding level:
$10,000

Company/organization:
Abbott Laboratories
Dept. 38-L Building A-11401 Sheridan Rd.
North Chicago, IL 60064

Contact name:
Gwendolyn Platt

Title:
Manager, Local Government Affairs

Telephone:
(708) 937-5107

45

ACTIVITY:
Tutoring

Program:
Home Instruction Program for Preschool Youngsters (HIPPY)Program objectives:Promote school readiness by enhancing parents' involvement in their child's academic success.

Target group:
Preschool

Year established:
1986

Program descriptions:
The Home Instruction Program for Preschool Youngsters (HIPPY) is a home-based school-readiness program for children at risk of failing in school. HIPPY is designed for parents with limited formal schooling to provide educational enrichments for their preschool children.

Parents are provided with daily instructional packets. They are required to work with their children fifteen minutes a day, five days a week, thirty weeks a year for two years, the second of which is the year the child is in kindergarten. The parent is trained by a paraprofessional from the same community who also has a four-year old in HIPPY. The paraprofessional visits the home every week. Twice a month the parents gather to share their experiences with their peers and to participate in enrichment programs covering topics such as health, safety, stress management, and job training.

HIPPY was introduced to the U.S. in 1986 by First Lady Hillary Rodham Clinton. Then First Lady of Arkansas, she had read about HIPPY, a program developed in Israel, in the *Miami Herald* and decided to investigate the prospects of bringing it to

Arkansas. She became its advocate in Arkansas, where the first pilot programs were launched in 1986. From 1986 to 1990 HIPPY programs were coordinated at the governor's office. In 1991 the Arkansas Better Chance Bill was signed, which provided $2.5 million to fund thirty-three HIPPY sites in Arkansas. There are sixty HIPPY programs in seventeen states, reaching approximately 10,000 disadvantaged families. All HIPPY programs in the U.S. receive training and technical assistance through HIPPY USA. Businesses provide funding for local sites.

Annual funding level:
Program costs average $1,000 per family per year over two years. Funding is provided at the federal, state, and local levels as well as from corporations, foundations, and community organizations.

Company/organization:
HIPPY USA
53 W. 23rd St.
New York, NY 10010

Contact name:
Kathryn Greenberg

Title:
Community Outreach Coordinator

Telephone:
(212) 645-2006

46

ACTIVITY:
Tutoring

Program:
Homework Hotline

Program objectives:
Emphasize the importance of school and learning, aid students in developing effective problem-solving strategies, and better equip parents to help their children with assignments.

Target group:
Kindergarten through twelfth grade

Year established:
1984

Program descriptions:
The Homework Hotline allows students from kindergarten to twelfth grade and their parents in Missouri and Illinois to call the toll-free hotline about homework problems in all subject areas, including math, English, social studies, science, and reading. The program is sponsored by the Missouri National Education Association (MNEA), McDonnell Douglas, Monsanto and KDNL Fox 30. It is a partnership of the public/private sector with teachers and professional engineers, scientists, and computer specialists from two major corporations collaborating in a joint effort.

More than 100 volunteers, working from a large conference room at the McDonnell Douglas facility in St. Louis, operate the dozen phone lines four nights a week, Monday through Thursday from 6 to 9 P.M. The program runs for nine months each year, from after Labor Day to Memorial Day. Volunteers have access to a library of textbooks, teacher editions, and other reference materials on site to use in working with their callers.

McDonnell Douglas began its hotline nine years ago working with just one high school. In 1991-92 the McDonnell Douglas Homework Hotline merged with the Homework Hotline sponsored by the National Education Association (NEA) and now serves over 500,000 students from more than 115 school districts in Missouri and Illinois. Volunteers fielded 16,000 calls in the past year. The NEA promotes the program to school districts, and KDNL-Fox 30 provides free public service announcements during its children's programming. The program is the largest hotline in the nation.

Annual funding level:
$40,000: toll-free lines are $20,000 to 25,000 per year; promotional material cost approximately $15,000.

Awards:
Bellringer Award of Merit from the Community Relations Report.

Company/organization:
McDonnell Douglas Corporation
Mailcode 1001530
P.O. Box 516
St. Louis, MO 63166-0516

Contact name:
Bonnie Brandt

Title:
Manager, Community Relations

Telephone:
(314) 232-8020

47

ACTIVITY:
Tutoring

Program:
The Junior Great Books Read-Aloud Program

Program objectives:
Higher literacy for all students.

Target group:
Kindergarten and first grade

Year established:
1990

Program descriptions:
The Junior Read-Aloud Program introduces prereaders and early readers to the pleasure of good literature and the experience of intellectual exchange. Children listen as an adult reads a story or poem to them and express their interpretations through drawing, dramatization, and sharing questions and ideas. The stories and poems in the Read-Aloud series are selected from outstanding traditional and modern literature from cultures around the world. Student books, activity books, teacher's editions, and leader's guides and support materials such as bookmarks, posters, and awards are available. The Basic Leader Training Course, a two-day intensive workshop taught by a Foundation instructor, is required for those who plan to lead a Junior Great Books program.

Schools that implement Junior Great Books programs as part of their regular curriculum can usually acquire district funds to implement the program. Federal chapter funds and school improvement funds are often available for special groups of students. Community organizations, parent/teacher organizations, and local businesses can provide funding and volunteers for the Read-Aloud program and materials.

The Great Books Foundation also offers reading and discussion programs for children in the second to twelfth grades and for adults. Approximately twenty percent of the school districts in the country use the Junior Great Books Read-Aloud Program.

Publications:
Read-Aloud, Junior Great Books and Great Books Reading Series.

Annual funding level:
Basic Leader Training Course is $70; curriculum training and classroom material costs vary.

Company/organization:
The Great Books Foundation
35 E. Wacker Dr., Suite 2300
Chicago, IL 60601-2298

Contact name:
Steven N. Craig

Title:
Editor

Telephone:
(312) 332-5870

48

ACTIVITY:
Tutoring

Program:
My Friend Taught Me

Program Objectives:
To increase Burns School children's academic skills, self-esteem, and motivation for learning by one-on-one tutoring.

Target group:
Grades 4-6

Year Established:
1989

Program Description:
Over 400 CIGNA employees participate in "My Friend Taught Me," a one-on-one tutoring program for at-risk fourth to sixth graders from the Dominick F. Burns School in Hartford. The mission of the program is to offer support and encouragement to the Burns School students, help them improve their academic skills, and expose them to an environment that is different from their life in Hartford's inner city.

Every Wednesday afternoon students are bussed to three CIGNA locations where employees spend an hour a week tutoring students in math and reading. Tutors have the choice of working alone or in teams. Tutors receive lesson plans each week to reflect what the children are learning in their classrooms. The focus is on helping the students with their homework and improving their communications skills. Special events, such as spelling bees, field trips and guest-speakers are held throughout the year. Students who are selected to be in "My Friend Taught Me" usually have a combination of special needs ranging from academic to social. English is a second language for most students at Burns, where 97 percent of the students are from minorities. To qualify for the program the students must be able to communicate in English and exhibit good behavior at school. The program began in 1989 with thirty fourth graders and has grown to involve 188 fourth, fifth and six graders. There are CIGNA coordinators who organize the program at each location and a full-time program manager who acts as the liaison to the school.

Annual funding level:
$70,000

Company/Organization:
CIGNA Corporation
900 Cottage Grove Road
Bloomfield, CT 06002

Contact Name:
Kathy Brady

Title:
Program Coordinator

Telephone:
(203) 726-5611

49

ACTIVITY:
Tutoring

Program name:
Project LIVE (Learning Through Industry and Volunteer Educators)

Program objectives:
Literacy and dropout prevention

Year established:
1972

Target group:
Seventh and eighth grade

Program descriptions:
Project LIVE (Learning Through Industry and Volunteer Educators) is a tutoring program for junior high school (seventh and eighth graders) in New York-area public schools who are two to three years below grade level. The program, developed by the Children's Aid Society, brings together schools, corporations, and social service organizations to tackle the problems of school dropouts and literacy. More than 250 volunteers from eight corporations participate in the program.

The Reader's Digest Association has sponsored Project LIVE for the past fifteen years. Students from Fox Lane Middle School in Bedford Hills, New York, are bussed every Tuesday afternoon to the Reader's Digest, Inc. offices. The volunteers work one-on-one with the students from 2:45 to 5 P.M. in a conference room. The students are encouraged to stay in the program for two years. The program runs from November through June. Volunteers are solicited each September through the in-house newsletter. Three two-hour training sessions led by Project LIVE trainers are held for the tutors at Reader's Digest. There are twenty-nine volunteers from all divisions of Reader's Digest

Condensed Books in the program. The Reader's Digest program coordinator works cooperatively with the school coordinator and the Children's Aid Society.

Program costs cover transportation, snacks for the students, the on-site tutoring library, and the salary of the liaison teacher who accompanies the students to the corporate site. The Children's Aid Society manages Project LIVE, providing the organizational and educational support to the corporate sponsor. The corporate sponsors provide the tutoring facilities, give their employees release time to participate in the program, and provide financial support.

Annual funding level:
Administrative and training costs available upon request from the Children's Aid Society.

Participating company/organization:
Reader's Digest Association
Reader's Digest Road
Pleasantville, NY 05170

Contact name:
Hazel Brown

Title:
Manager of Reprint Department

Telephone:
(914) 241-5373Program headquarters:
Children's Aid Society105 E. 22nd St.New York, NY 10010

Contact name:
Martha Cameron

Title:
Director of Youth Development Services

Telephone:
(212) 949-4800

50

ACTIVITY:
Tutoring

Program:
Time to Read

Program objectives:
Combat functional illiteracy and improve student's reading, thinking, and vocabulary skills.

Target group:
Adolescents age twelve years and older who are reading at fourth- to eighth- grade level.

Year established:
1985.

Program descriptions:
Time to Read is Time Warner's nationwide volunteer tutorial program designed to combat functional illiteracy. Tutors are trained to use Time Warner magazines, comics, music videos, and an activity-based curriculum to help learners enjoy reading, increase their vocabulary skills, and develop life-long reading strategies.

Tutors spend two hours a week for one year with one learner or a group of two to five learners. Independent evaluation surveys, completed every two years, show that seventy-nine percent of the learners improve their reading scores. The program now operates at 161 program locations in eighteen states and the District of Columbia.

Time Warner provides its subsidiaries with materials and training. The subsidiaries assume local costs for the program which may include busing the students, snacks, and recognition awards. Approximately forty percent of the 2,000 volunteers are from Time Warner subsidiaries.

Other sponsors, including community organizations, schools and local businesses, pay $175 per learner per year for on-site training and materials. The program has a full-time manager and an educational consultant.

Annual funding level:
$800,000

Awards:
1991 Council on Economic Priorities (CEP) Corporate Conscience Award, 1988 President's Volunteer Action Award.

Participating company/organization:
Time Warner, Inc.
1271 Avenue of the Americas
New York, NY 10020

Contact name:
Toni Fay

Title:
Director, Corporate Community Relations

Telephone:
(212) 522-1485Program headquarters:Time to Read1271 Avenue of the AmericasNew York, NY 10020

Telephone:
(212) 522-6917
See page 44 for success story.

RESOURCE DIRECTORY

The following national and local organizations can assist businesses and individuals who want to become involved in school/business partnerships and education reform as program sponsors, funders, volunteers, or advocates. These organizations conduct research, seminars, and training; lobby for educational reform; and sponsor and develop partnership programs.

Detailed information on the services offered by each organization can be obtained directly by calling or writing to the names and addresses listed below.

Advocacy/public policy/research

Business Roundtable
1615 L St. NW, Suite 1350
Washington, DC 20036
(202) 872-1260
Christopher Cross, Executive Director, Education Initiatives

Committee for Economic Development
477 Madison Ave.
New York, NY 10022
(212) 688-2063
Sandra Hamburg, Vice President & Director of Education Studies

Conference Board
Business/Education Council
845 Third Ave.New York, NY 10022
(212) 759-0900
Leonard Lund, Manager

Education Commission of the States
707 17th St., Suite 2700
Denver, CO 80202
(303) 299-3611
Frank Newman, President

National Alliance of Business
Center for Excellence in Education
1201 New York Ave. NW
Washington, DC 20005
(202) 289-2925
Esther Schaeffer, Senior Vice President

National Center on Education and the Economy
39 State St., Suite 500
Rochester, NY 14614
(716) 546-7620
Mutui Fagbayi, Chief Operating Officer

Public Education Fund Network
601 13 St. NW, Suite 370S
Washington, DC 20005
(202) 628-7460
Wendy Puriefoy, President

Business/school partnerships

Business Higher Education Forum
One Dupont Circle, Suite 800
Washington, DC 20036
(202) 939-9345
Judith Irwin, Associate Director

Council for Aid to Education
51 Madison Ave., Suite 2200
New York, NY 10010
(212) 689-2400
Diana W. Rigden, Director of Precollege Programs

National Association of Partners in Education, Inc.
209 Madison St., Suite 401
Alexandria, VA 22314
(703) 836-4880
Daniel W. Merenda, President & CEO

National Dropout Prevention Center
Clemson UniversityClemson, SC 29364
(803) 656-2599
Jay Smink, Executive Director

Points of Light Foundation
736 Jackson Place
Washington, DC 20503
(202) 408-5162
Patricia Bland, Coordinating Vice President for Leadership
and Product Development

U.S. Department of Education
Office of Educational Research and Improvement
Washington, DC 20208
(202) 219-2116
Sue Gushkin, Coordinator of Educational Partnerships

Education organizations

Council for Basic Education
725 15th St. NWWashington, DC 20005
(202) 347-4171
A. Graham Down, President

National Committee for Citizens in Education
900 Second St. NE, Suite 8
Washington, DC 20002
(202) 408-0447
Bruce Astrein, Executive Director

Council of Chief State School Officers
1 Massachusetts Ave. NW
Washington, DC 20001
(202) 408-5505
Glenda Partee, Assistant Director

National Community Education Association
119 N. Payne St.Alexandria, VA 22314
(703) 683-6232
Starla Jewell-Kelly, Executive Director

National Education Association
1201 16th St. NWWashington, DC 20036
(202) 833-4000
Keith B. Geiger, President

National Parent Teacher Association
700 N. Rush St.Chicago, IL 60611
(312) 878-0977
Pat Henry, President

Literacy

The Center for the Book in the Library of Congress
Washington, DC 20540
(202) 707-5221
John Cole, Director

Coalition for Literacy
50 E. Huron St.Chicago, IL 60611
(312) 944-6780
Mattye Nelson, Staff Liaison, American Library Association

Laubach Literacy Action
5795 Widewaters Pkwy.Syracuse, NY 13214
(315) 422-9121
Peter Waite, Executive Director

Literacy Volunteers of America
5795 Widewaters Pkwy.Syracuse, NY 13214
(315) 445-8000
Helen Crouch, President

Reading Is Fundamental
600 Maryland Ave. SW, Suite 500
Washington, DC 20560
(202) 287-3220
Ruth Graves, President

School reform/restructuring

Accelerated Schools Project
CERAS BuildingStanford University
Stanford, CA 94305
(415) 723-0840
Henry Lewis, Director

Center for Collaborative Education
1573 Madison Ave., Room 201
New York, NY 10029
(212) 348-7821
Heather Lewis & Priscilla Ellington, Co-directors

Child Study Center
Yale University
School Development Program
230 S. Frontage Rd.P.O. Box 3333
New Haven, CT 06510
(203) 785-2548
James P. Comer, Director

Coalition for Essential Schools
Brown University
Box 1969
Providence, RI 02912
(401) 863-3384
Theodore Sizer, Chairman

New American Schools Development Corporation
100 Walnut Blvd., Suite 2710
Arlington, VA 22209
(703) 908-9500
W. Frank Blount, President

New York City resources

Mayor's Office of Education Services
52 Chambers St.New York, NY 10007
(212) 788-3271
Robert Steinman, Policy Analyst

New York Alliance for the Public Schools
32 Washington Pl., Fifth Floor
New York, NY 10003
(212) 753-3090
Barbara Pobst, Executive Director

New York City Board of Education
Office of External Programs
110 Livingston St.Brooklyn, NY 11201
(718) 935-5311
Clare O'Connor, Director

New York City School Volunteer Program, Inc.
443 Park Ave. South
New York, NY 100016
(212) 213-3370
Susan Edgar, Executive Director

Mayor's Voluntary Action Center
61 Chambers St.New York, NY 10007
(212) 566-5950
Winifred Brown, Director

Public Education Association
39 W. 32nd St.New York, NY 10001
(212) 868-1640
Jeanne Silver Frankl, Executive Director

The Governor's School and Business Alliance
11 W. 42nd St., 21st Floor
New York, NY 10036
(212) 790-2490
Kristen O. Murtaugh, Executive Director

United Parents Association of NYC, Inc.
45 John St., Suite 607
New York, NY 10038
(212) 406-7068
Jan Atwell

Fund for New York City Public Education
96 Morton St., Ninth Floor
New York, NY 10014
(212) 645-5110
Beth Lief, Executive Director

New York Mentoring School and Business Alliance
267 Fifth Ave., Suite 1003
New York, NY 10016
(212) 779-3620
Marjorie Wilkes, Executive Director

NATIONAL ASSOCIATION OF PARTNERS IN EDUCATION CONTACTS BY STATE

AZ NAPE
Dr. Charles Hoyt
President
Arizona Alliance for Math & Science
Phoenix, AZ

California School Volunteer Program
Kay Bergdahl
Community Coordinator
Edison High School
Huntington Beach, CA

Colorado Association of Partners in Education
Tara Zeleny
District Volunteer Coordinator
Fort Collins, CO

Connecticut Association of Partners in Education
Leroy Spiller
Chair, Council of State Affiliate Presidents
East Hartford, CT

D.C. Association for Partnerships in Education
Audrey Epperson
Director, Education Services
Washington Urban League
Washington, DC

Florida Association of Partners in Education
Marge Baker
Alachua County Public Schools
Gainesville, FL

Georgia Partners in Education
Anna Burns
Director, Future Stock
Griffin-Spalding School System
Griffin, GA

Illinois School Volunteer and Partnership Programs
Barbara Banker
Director, Community Services
Woodstock Unit School District 200
Woodstock, IL

Iowa School Volunteer Network
Sue Pearson
Cedar Rapids Community School District
Cedar Rapids, IA

Louisiana Association of Partners in Education
Joseph L. Hebert
Vice President of Administration

Central Louisiana Chamber of Commerce
Alexandria, LA

Maine Alliance of Partners in Education
Sandra Schniepp
Chairperson
Maine Alliance of Partners in Education
Kingfield, ME

Maryland Association of Partners in Education
Paula Blake
Assistant to the Superintendent of Schools
Howard County Department of Education
Ellicott City, MD

Massachusetts Association of Partners in Education, Inc.
Jane Schroeder
Volunteer Coordinator
School Volunteers for Milford
Milford, MA

Michigan School Volunteer Program
Richard Njus
Hiawatha Elementary School
Okemos, MI

National Association of Partners in Education/MN
Lois Norby
Coordinator of Volunteer Services
Minnetonka Public Schools
Excelsior, MN

New Hampshire Partners in Education
Nancy Craig
Executive Director

New Hampshire Partners in Education
Manchester, NH

New Jersey Association of Partners in Education
Charlotte Frank
Vice President, Research & Development
MacMillan-McGraw Hill School Pub. Co.
New York, NY

North Carolina Association of Volunteers & Partners in Education
Jon Allen
Partnership Coordinator
Granville County Schools
Oxford, NC

Ohio School Volunteer Partners, Inc.
Nancy White
Community Relations Coordinator
Dayton Public Schools
Dayton, OH

Oklahoma Association of Partners in Education
Judy Kokesh
Education Initiative Project
Department of Energy - NIPER
Bartlesville, OK

Oregon Community Education Association
Kathy Schrock
Director, Volunteer Development
Albany Public Schools
Albany, OR

Ocean State Association of Partners in Education
Lynn C. Murphy
Director
CHILD, Inc.
Warwick, RI

South Carolina Association of Partners in Education
Dr. Muriel O'Tuel
Conway, SC

Tennessee Association of Partners in Education
Barbara Russell
Director, Adopt-A-School Program
Memphis City Schools
Memphis, TN

Texas Association of Partners in Education
Kay Metz
Coordinator of School/Community Relations
Waco Independent School District
Waco, TX

Utah Association of Partners in Education
Daphne Williams
Coordinator of Volunteer Services
Salt Lake City Public Schools
Salt Lake City, UT

Vermont Chamber of Commerce Business/Education Partnership, Inc.
Winton Goodrich
Executive Director, Vermont Chamber of Comerce
    Business/Education Partnership
Montpelier, VT

Virginia Association of Partners in Education
Donna P. Caudill
Director, Business/Industry Relations
Fairfax County Public Schools
Falls Church, VA

Washington Association of Partners in Education
Barbara Schneider
Director of Community Relations
North Thurston School District
Lacey, WA

INDEXES

PROGRAM/FOUNDATION NAME PROFILE

Academy of Finance 9
Adopt-a-Student 23
Art and Writing Awards 38
BOOK IT! 34
Burger King Academy Career 29
Awareness in Middle Schools (CAMS) 10
Career Awareness Program(CAP) 16
Career Beginnings 17
Children of the Future 44
Children's Aid Society 49
CHOICES 11
Cities in Schools 29
Cray Academy 40
Editors in the Classroom 41
Everybody Wins 24
Hands on Science Project 33

Home Instruction Program for Preschool Youngsters (HIPPY) 45

Homework Hotline 46

HOSTS (Help One Student to Succeed) 251

Have a Dream 36

Junior Great Books Read-Aloud Program 47

Kaleidoscope Educational Series 32

Mayor's Summer Jobs Program 18

Middle School Attendance Incentive Program 37

My Friend Taught Me 48

National Academy Foundation 9

Partners in Excellence 4

Project LIVE (Learning through Industry and Volunteer Educators) 49

Project SEARCH (Seeking Experience: A Real Career Help) 14

Project STEP (Skills Training Educational Program) 19

Reading Is Recreation 1

Saturday Academy 20

Science and Mathematics (SAM) Mentoring Program 26

Shadow Day 21

Technical Teams Encouraging Career Horizons (TECH) 8

Thumbs Up 39

Time to Read 50

U S West Foundation 11

World of Work 22, 27

Youth Motivation Program 15

TYPE OF ACTIVITY PROFILE

Assemblies 32

Career days 2, 4, 5, 14, 31

Curriculum development/enhancement 22, 28, 30, 31, 33

Demonstrators 5, 32, 33

Donations 1, 4, 5, 7, 18, 26, 31

Employment 18

Fair 13

Field Trips 9, 16, 19, 20, 21

Grants 7, 26

Guest speakers 1, 5, 19, 20, 21, 32, 33

Incentives 2, 7, 29

In-kind donations 6

Internships 12, 28, 35

Job Training 16, 19

Job Shadowing 9, 14, 19, 20, 21

Leadership Conferences/Training 13, 28, 31, 40

Mentoring 4, 6, 9, 13, 17, 22, 23, 24, 25, 26, 27, 28, 29, 35

Physical plant improvement 7

Plant/business tours/visits 2, 4, 35, 40

Scholarships 5, 6, 9, 14, 16, 28, 35, 36

Speakers 7, 8, 9, 10, 11, 12, 15, 26

Special awards 1, 5, 34, 37, 38, 39

Special events 2, 3, 12, 32

Staff development/training 28, 41

Student employment 2, 29

Summer Internships 9

Teacher training/workshops/conferences 9, 30, 41, 42

Tours 5, 10, 35

Tutoring 1, 2, 3, 4, 5, 6, 19, 21, 22, 28, 29, 31, 43, 44, 45, 46, 47, 48, 49, 50

Workshops 18, 31, 35, 40

COMPANY/ORGANIZATION PROFILE

Abbott Laboratories 44

Accelerated Schools Project Resource Directory 175

Aetna Life & Casualty Corp. 20

Bank of America 19

American Express 9

Bantam Doubleday Dell 1

Matthew Bender & Company 18

Boeing Company 43

Burger King 29

Business Higher Education Forum
    Resource Directory 173

Business Roundtable Resource Directory 171

Cahners Publishing Company 27

Center for Collaborative Education
    Resource Directory 176

Center for the Book in the Library of Congress
    Resource Directory 175

Champion International Corporation 30

Chevron Chemical Company 21

Child Study Center Resource Directory 176

Chrysler Corporation 22

Cigna Corporation 48

Coalition for Essential Schools Resource Directory 176

Coalition for Literacy Resource Directory 175

Columbia Regional Hospital 14

Committee for Economic Development Resource
    Directory 172

Conference Board Resource Directory 172
Council for Aid to Education Resource Directory 173
Council for Basic Education Resource Directory 174
Council of Chief State School Officers
    Resource Directory 174
Cray Research, Inc. 40
Dow Chemical Company 33
Eastman Kodak Company 28
Education Commission of the States
    Resource Directory 172
Fannie Mae 35
Fund for New York City Public Education
    Resource Directory 178
General Motors, Harrison Division 12
The Great Books Foundation 47
HIPPY (Home Instruction Program for Preschool
    Youngsters) USA 45
Home Savings of America 16
HOSTS (Help One Student to Succeed) Corp. 25
Houghton Mifflin Company 41
I Have a Dream Foundation 36
Ingram Distribution Group, Inc. 13
Laubach Literacy Action Resource Directory 175
Literacy Volunteers of America Resource Directory 175
Macmillan/McGraw-Hill School Publishing Company 24
McDonald's Family Restaurants 37
McDonnell Douglas Corp. 46
Mobil Oil Corporation 2
Monsanto 25
Motorola, Inc. 31
National Academy Foundation 9
National Alliance of Business Resource Directory 172
National Association of Partners in Education Resource
    Directory 173

National Center on Education and the Economy
    Resource Directory 172

National Committee for Citizens in Education Resource
    Directory 174

National Community Education Association Resource
    Directory 174

National Dropout Prevention Center Resource
    Directory 173

National Education Association Resource Directory 174

National Parent Teacher Association
    Resource Directory 174

New American Schools Development Corporation
    Resource Directory 176

New York Alliance for the Public Schools Resource
    Directory 177

New York City Board of Education Resource
    Directory 177

New York City Mayors Office of Education Services
    Resource Directory 176

New York City Mayors Voluntary Action Center Resource
    Directory 177

New York City Public Education Association Resource
    Directory 178

New York City School Volunteer Program Resource
    Directory 177

New York Mentoring School and Business Alliance
    Resource Directory 178

New York State—The Governor's School and Business
    Alliance Resource Directory 177

Penguin USA 3

Pizza Hut 34

Points of Light Foundation Resource Directory 173

Proctor & Gamble Paper Products Co. 4

Public Education Fund Network Resource Directory 172

Reader's Digest Association 49

Reading is Fundamental Resource Directory 175
SC Johnson Wax 32
Scholastic, Inc. 38
Scott, Foresman & Company 15
Springfield Institution for Savings 23
Southern Bell 11
Sunbank/Miami N.A. 39
Tenneco, Inc. 6
Texas Instruments 42
Thom McAn 17
Time Warner, Inc. 50
Toyoto, T.A.B.C., Inc. 5
The New England 73
M 8
UNISYS Corporation 10
U.S. Department of Education Resource Directory 173
United Parents Association of NYC
    Resource Directory 178
Westinghouse Electric Corporation 26

Those companies/organizations with no profile number appear on the Resource Directory, pages 171-178.